fiona Maclennan

*Polycrystalline and Amorphous
Thin Films and Devices*

MATERIALS SCIENCE AND TECHNOLOGY

EDITOR

A. S. NOWICK

Henry Krumb School of Mines
Columbia University
New York, New York

Polycrystalline and Amorphous Thin Films and Devices

Edited by

LAWRENCE L. KAZMERSKI

Photovoltaics Branch
Solar Energy Research Institute
Golden, Colorado

 1980

ACADEMIC PRESS
A Subsidiary of Harcourt Brace Jovanovich, Publishers

New York London Toronto Sydney San Francisco

ACADEMIC PRESS, INC.
111 Fifth Avenue, New York, New York 10003

United Kingdom Edition published by
ACADEMIC PRESS, INC. (LONDON) LTD.
24/28 Oval Road, London NW1 7DX

Library of Congress Cataloging in Publication Data
Main entry under title:

Polycrystalline and amorphous thin films and devices.

    (Materials science and technology series)
    Bibliography: p.
    Includes index.
    1. Thin films. 2. Thin film devices. 3. Semi-
conductor films. 4. Photoelectronic devices. 5. Optical
films. 6. Protective coatings. I. Kazmerski, Lawrence L.
TK7871.15.F5P647   621.381'73   79–8860
ISBN 0–12–403880–8

PRINTED IN THE UNITED STATES OF AMERICA

80 81 82 83   9 8 7 6 5 4 3 2 1

To my mother, my wife Kathleen,
and to Keira and Timothy

Contents

List of Contributors

Numbers in parentheses indicate the pages on which the authors' contributions begin.

ALLEN M. BARNETT* (229), *Institute of Energy Conversion, University of Delaware, Newark, Delaware 19711*

PATRICK J. CALL (257), *Solar Energy Research Institute, Golden, Colorado 80401*

D. E. CARLSON (175), *RCA Laboratories, Princeton, New Jersey 08540*

A. H. CLARK (135), *University of Maine at Orono, Orono, Maine 04469*

DAVID EMIN (17), *Division 5151, Sandia National Laboratories, Albuquerque, New Mexico 87185*

LEWIS M. FRAAS (153), *Chevron Research Company, Richmond, California 94802*

LAWRENCE L. KAZMERSKI (59), *Photovoltaics Branch, Solar Energy Research Institute, Golden, Colorado 80401*

BRENTON L. MATTES (1), *University of Michigan, Ann Arbor, Michigan 48109*

JOHN D. MEAKIN (229), *Institute of Energy Conversion, University of Delaware, Newark, Delaware 19711*

R. A. MICKELSEN (209), *The Boeing Aerospace Company, Seattle, Washington 98124*

ALLEN ROTHWARF† (229), *Institute of Energy Conversion, University of Delaware, Newark, Delaware 19711*

KENNETH ZANIO (153), *Hughes Research Laboratories, Malibu, California 90265*

*Present address: Electrical Engineering Department, University of Delaware, Newark, Delaware 19711

†Present address: Electrical Engineering Department, Drexel University, Philadelphia, Pennsylvania 19104

Preface

Considerable efforts have been expended in trying to understand and predict the electronic, optical, and physical properties of thin material layers. Remarkable progress has been made in these areas, interrelating the basic characteristics and in some instances, applying the results in the realization of active and passive device technologies. Thin films of various materials are currently in use as protective and optical coatings, selective membranes, and thermal transfer layers. Metal, semiconductor, and insulator thin films are used extensively in the electronics industry, with applications ranging from submicron area very large-scale integration (VLSI) memory units to large-areal (comprising many square miles) energy conversion devices. Although a number of excellent reviews and books on this subject are found in the literature (see the bibliography at the end of the book), most of them focus on the behavior of metal and insulator thin films, dealing with the semiconductor film in a more superficial manner. Semiconductor thin film R & D activity is evolving dramatically, following intense interest in low-cost, large-scale applications. However, many of the major contributions in this expanding field remain segmented in the literature. It is the purpose of this book to consolidate this information in a single source that can provide a general basis for understanding polycrystalline and amorphous semiconductor thin films and devices.

This book is organized into two parts. The *first* (Chapters 1–5) deals with the basic properties—growth, structure, electrical and optical mechanisms—encountered in amorphous and polycrystalline thin semiconductor films. The *second* part (Chapters 6–9) covers the applications and problems of these layers in active semiconductor devices and passive technologies. The authors were chosen primarily on the basis of their research activities in the selected topical areas, and they represent university, industrial, and national research laboratories.

The book begins with an introduction to the several mechanisms that suggest a hierarchy in the growth and structure (solid, liquid, and vapor phases) of amorphous and polycrystalline films [Brenton L. Mattes, University of Michigan]. This is followed by a synopsis of the electrical and optical properties of *amorphous* thin films [David Emin, Sandia National Laboratories]. This chapter discusses the structure of covalently bonded amorphous semiconductors, provides various models of localization, contrasts the amorphous semiconductor during and after illumination, and considers the important issues confronting research in this field. The electrical properties of *polycrystalline* semiconductor thin films are reviewed in Chapter 3 [Lawrence L. Kazmerski, Solar Energy Research Institute]. This treatment focuses on the transport phenomena in elemental and compound semiconductor films, with some special emphasis placed on grain boundary contributions. The optical properties of these films, primarily dealing with the optical constants relating to electronic structure, are covered in Chapter 4 [Alton H. Clark, University of Maine]. The discussion of basic mechanisms topics concludes with Chapter 5, which details the electronic structure of grain boundaries in polycrystalline semiconductors [Lewis M. Fraas, Chevron Research Company, and Kenneth Zanio, Hughes Research Laboratories]. A key discussion of the methods of defect modification leading to grain boundary passivation is included. The applications topics begin with a thorough examination of active *amorphous* thin-film devices [David E. Carlson, RCA Laboratories]. A review of amorphous device technology, experimental methods, recent developments in thin-film devices based on hydrogenated amorphous silicon, and predictions of future directions for this technology are covered. Devices based on *polycrystalline* semiconductor thin films are summarized in Chapter 7 [Reid A. Mickelsen, The Boeing Aerospace Company]. Thin-film transistors, diodes, photoconductors, and luminescent films form the basis for this chapter. A future large-scale, large-area device, the thin-film solar cell, is overviewed in Chapter 8 [Allen Rothwarf, John D. Meakin, and Allen M. Barnett, Institute of Energy Conversion, University of Delaware]. The present research situation for these promising devices is discussed, and forecasts of future technologies are presented. Finally, the important area that addresses the applications of *passive* thin films, including their functions, materials selection, and taxonomy, is treated [Patrick J. Call, Solar Energy Research Institute]. Topics include optical films, protective coatings, corrosion, high and low energy surfaces, and selective membranes.

The understanding of the basic mechanisms that control the processes and properties of thin semiconductor films continues to evolve. The applications of these films expand as more ideas are generated. Device improvement and diversity persist. The future of the semiconductor thin film, however, depends on the ingenuity and expertise of those involved in the research and development activities. It is the sincere hope of the editor and authors that this book can especially serve that group and aid in the future deployment of technologies based on amorphous and polycrystalline thin films.

The editor expresses his sincere appreciation to Sigurd Wagner for his suggestions and encouragement in organizing the book. The valuable assistance of Peter Sheldon and Phillip J. Ireland in editing and reviewing the manuscripts is gratefully acknowledged. Finally, the editor wishes to recognize and to thank Susan Sczepanski who provided (with great diligence) many of the figures in the book, and Betsy Fay-Saxon who prepared (with great patience) the manuscripts into their final formats.

1 | Growth and Structure of Amorphous and Polycrystalline Thin Films

BRENTON L. MATTES

University of Michigan
Ann Arbor, Michigan

1.1 INTRODUCTION

The mechanisms involved in the formation of crystalline or noncrystalline states by condensation from vapor and liquid phases primarily depend on the time that atoms or clusters of atoms interact to form bonds in metastable and stable structures. Crystallization is the long-range ordering of atoms in a periodic solid-phase lattice near equilibrium conditions. There are many comprehensive treatments on the nucleation and growth of crystalline solids [1–7] and thin films [8–12]. Basically, the theories assume adatom attachment and/or phenomenological thermodynamic and Arrhenius relationships [6, 7, 11, 12]. Although theories on crystallization appear to be at hand, there are few theoretical treatments on the formation

1

of the amorphous solid state. Only qualitative thermodynamic and energetic driving forces have been proposed [7, 13, 14]. However, there are many detailed studies on structural models [15–26], deposition processes [27–31], and physical properties [16, 27, 28, 33–39].

This review will not necessarily fill the gap but is intended to introduce several mechanisms that seem to suggest a hierarchy in the growth and structure of condensed states of the solid, liquid, and vapor phases.

1.2 REVIEW OF AMORPHOUS AND POLYCRYSTALLINE STATES

Amorphous and, in general, polycrystalline thin films possess no unique directionality or axis on a macroscopic plane. Microscopically, there may be some debate as to whether an amorphous state with only short-range order is a random dense packing of *atoms* [13, 17, 24] or *microcrystallites* (clusters 10–15 Å in diameter) [15, 16, 23, 24], as either may exist in the liquid phase. Polycrystalline materials are crystalline but with random grain size, shape, and orientational packing.

1. Amorphous State

The amorphous state appears to require bonding anisotropies associated with the polymorphisms of elemental solids [40] and, in addition, atom size differences for alloy and compound solids [41]. The most general process used to form amorphous materials is to quench from the liquid phase. This in general only works for the elements S, Se, P, As, and B (Fig. 1.1, region I) [40] and for metallic glasses* such as $Pd_{0.8}Si_{0.2}$ [21, 42]. Other elemental amorphous materials are obtained by a variety of vapor, electrolytic, and sputter processes that deposit materials onto substrates at different temperatures. In addition, ion implantation can produce an amorphous state by creating locally vapor-quenched pockets in a surface layer of crystalline material [24].

As polymorphism decreases (Fig. 1.1, region II), the substrate temperatures must be below $\sim 400°C$ for C, Si, and Ge, and < 77 K for Ni, Fe, Bi, Sb, and Te. Elements in region III can only be amorphous when the substrates are at ~ 4 K. In region IV, not only are very low substrate temperatures required, but stabilizing factors such as contamination, very thin films, and favorable substrate interactions are necessary. Only Na, K, Cs, and Rb with nearly isotropic bonding (region V) have not exhibited an amorphous state.

---

*Glasses are amorphous solids that might be described as being composed of two or more compounds with differing cluster sizes and/or cluster formation temperatures.

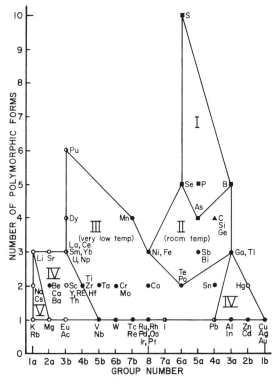

Fig. 1.1 A regional map showing the occurrence of elemental noncrystalline solids as correlated with the number of polymorphic forms and group number for all the solid elements in the periodic table. The boundary lines are assigned to group the elements into five regions representing the ease of forming noncrystalline solids: ■, quenching from the liquid; ▲, formation by chemical or physical vapor deposition near room temperature; ●, formation by vapor deposition at low temperatures; ◑, no data available but believed to form under corresponding conditions; and ○, no data available (after Wang and Merz, [40]).

Crystalline state bonds still predominate in amorphous solids. The thermodynamic stability for any crystalline form will be the bonding configurations at a given temperature with the lowest free energy. Rapid quenching from high-energy states in the liquid or vapor phase will prevent the atoms from relaxing into the lowest energy state. Therefore, elements with a large number of polymorphic forms will relax into a greater number of available bonding states [40]. This will lead to a random-packing network of atoms.

On the other hand, rapid quenching will also prevent atoms from attaching to embryos (unstable clusters) or nuclei (stable clusters) beyond a certain size as a result of kinetics and orientationally favorable growth directions. In addition, there may be several atom cluster sizes that are stable at different temperatures, e.g., molecules in the vapor phase, small atom clusters in the liquid and amorphous phases, and large atom clusters (20–50 Å in diameter) in the shape of close-packing polyhedra in the crystalline solid phase [43]. The cluster mechanism will be discussed in Section 1.5, in an attempt to interrelate all states found in the solid phase.

2. Polycrystalline State

To assess the polycrystalline state, with longer short-range order, the thermodynamic stability is not only dictated by the crystalline bonding configurations with the lowest free energy but also by the grain boundary energy, suface diffusion, and coalescence on the substrate. The grains or crystallites are formed by independent nucleation and growth processes randomly oriented and spaced with respect to one another. There is, in general, no epitaxial registration with the substrate lattice other than for nucleation occurring preferentially at defect sites and other surface irregularities. The grain size is controlled by the number of nucleating sites, which can be increased by rapid quenching, by seeding with additives, or with highly abraded substrate surfaces. Grain growth ceases when surrounding grains restrict further growth. Recrystallization may then be induced by annealing which reduces the grain boundary surface area by diffusion [1].

The shape of the crystallite is determined by the surface orientations favorable for adatom attachment from the solid, liquid, and vapor phases. This surface is determined by the orientation with the lower interfacial energies such as the {110} or {111} orientation for diamond cubic [44, 45] and zinc blende [46] and the {100} orientation for NaCl structures. Deposition of these materials will, in general, show a high degree of preferential orientation normal to the plane of the substrate.

1.3 QUENCHING PROCESS INTERACTIONS

The quenching process simultaneously involves the rapid cooling of a vapor or a liquid phase through polymorphic equilibrium transformation temperatures and the formation of crystalline bonds in a liquid- or solid-phase state, respectively. The cooling rate necessary to form an amorphous state must approach a rate determined by thermal fluctuations (latent heat or specific heat) at the transformation temperature that occurs when atoms and/or clusters are allowed to interact for a period of time. The maximum interaction time is approximately determined by the time it takes for bonds to form for short-range order over ~ 10 Å. Once bonds have been formed, lattice waves commence and propagate at 10^5 cm/s. Therefore, the maximum interaction time is $\sim 10^{-12}$ s, which is also close to the lattice vibration period (Debye period). If the magnitude of the thermal fluctuation or the range for crystallization is $\sim 1°C$, then the cooling rate will have to approach $10^{12}°C/s$ to obtain an amorphous state. However, allowing for heat conduction away from the adsorbed atom or cluster to the surroundings [46], the fluctuation may be $\sim 10^{-6}°C$, thus requiring cooling rates of $\sim 10^{-6}°C/s$.

Similar arguments [14] also hold for transfer of the kinetic energy of an incident vapor atom to a crystal lattice by estimating the mechanical relaxation time. The thermal fluctuation is then the temperature decrease experienced by the adsorbed atom.

1.4 GROWTH OF AMORPHOUS AND POLYCRYSTALLINE THIN FILMS

Two conclusions can be drawn based on the growth of amorphous and polycrystalline thin films: (1) The growth of either state is independent of the substrate (assuming no substrate–film interaction) whether it is amorphous or crystalline, and (2) an amorphous material can be transformed to a polycrystalline state, but not the reverse. The latter irreversibility indicates that the crystalline state has a lower lattice energy. Indeed, the polycrystalline state, too, will transform to a single-crystal state by the reduction of internal surface (grain boundary) energy. One might also infer that the transition from amorphous to polycrystalline state occurs by the reduction of internal surface energy. Note, however, that in both cases internal surfaces are reduced or minimized but not necessarily eliminated. Thus one could infer that single crystals also have internal surfaces, e.g., stacking faults. This conclusion is substantiated by the further observation that the degree of supercooling of a liquid increases with the degree of superheating since the crystal was last melted; i.e., the liquid retains a memory of the solid structure [47]. Energetically, melting and freezing involve a certain amount of latent heat which for metals is just about the energy required to melt or freeze a single layer of atoms [47]. The internal surface energy is about one-half of the latent heat. In addition, the energy of a lattice in shear distortion is comparable to the latent heat of a lattice on melting or freezing [48]. Similar observations have also been made on the vapor-to-solid phase transition where islands of atoms move around on a surface before fixing their position [49, 50].

The concept of single adatom adsorption and desorption at the surface of a critically sized embryo or a nucleus to describe growth and melting appears to be rather simplistic in that surface energies rather than single bond energies play the major role. Therefore, it is not surprising that the amorphous and polycrystalline states of condensed matter are independent of the substrate. One must therefore focus on the surfaces of small groups of atoms to understand the nucleation and growth processes at or ahead of the solid interface.

1.5 CLUSTER MODEL APPLIED TO Si AND AMORPHOUS Si:H ALLOYS

Based on experimental evidence, clusters have been detected and/or invoked to describe (1) metallic glasses formed at and near the eutectic composition above the eutectic temperature [42] and soda lime–Si glasses [41]; (2) the crystallization of alkali halides from aqueous solutions [51, 52]; (3) expansion of gases through conical nozzles [53–55]; (4) superlattice surface structures [43, 46] and thermal conductivity [56] in III–V compounds; and (5) Si vapors from both solid and liquid thermal sources [57]. These atom clusters range from several atoms in the vapor phase to ~ 1000 atoms in the liquid and solid phases. By deduction, clusters appear to transcend through the condensed phases of matter in some form of hierarchy. In order to describe this hierarchy, a cluster model will be presented that may represent the events that occur in the growth and lead to the structure of the solid states of Si.

1. Clusters

A cluster is an arrangement of atoms that bond crystallographically and have a unique stable (sometimes metastable) size and shape. Cluster size is determined by several factors: (1) a minimum lattice energy where bonds in surface tension balance bonds in bulk compression; (2) a close-packing polyhedral shape; (3) stoichiometry, if a compound; and (4) hybridized dangling bonds on the surface. In general, these conditions are simultaneously met for a stable cluster. Figure 1.2 shows Si clusters with chain- and solidlike forms. The transition to a solid cluster occurs for ten atoms arranged at the corners of a rhombic dodecahedron. This is the minimum number and the only polyhedral shape that will place the atoms in a single shell with tetrahedral coordination. The rhombic dodecahedron shell is a polyhedron with 12 {110} facets and close packs when the clusters are the same size. Shells of atoms are added until stable or metastable sizes are reached. There are several methods for determining lattice energies and/or force balances [46]. For Si a convenient method is to calculate the total energy contained in bonds which include the hybridized surface bonds (Fig. 1.3) when all the atoms are brought together to

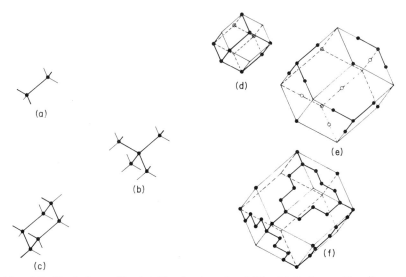

Fig. 1.2 Si clusters with chainlike (a–c) and solidlike (d–f) forms. Dangling and completed bonds in tetrahedral coordination are shown for the chainlike Si clusters. The outer rhombic dodecahedra shells (d–f) are for one-, two- and three-layered Si clusters, respectively.

form a cluster. The numbers of internal bonds and distributions of hybridized bonds on the surface for each cluster size are given in Table 1.1. The approximate energies associated with hybridized first, second, and third nearest-neighbor bonds, by stretching a Si—Si bond, are given in Table 1.2. The results for pure Si and for clusters with all dangling bonds bonded to H are shown in Table 1.3. It is important to note that this calculation

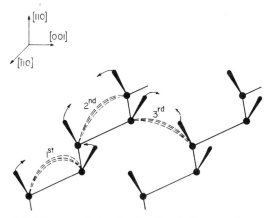

Fig. 1.3 The hybridization of the dangling bonds between first, second, and third nearest neighbors on a (110) Si surface.

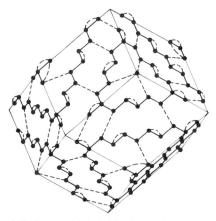

Fig. 1.4 A six-layered Si cluster with hybridized dangling surface bonds. Note that the third nearest-neighbor hybridized bonds provide the surface tension necessary to stabilize the cluster.

TABLE 1.1

Cluster Bond Distribution

Cluster size (no. of shells)	No. of Si atoms	No. of internal bonds	No. of hybridized surface bonds[a]		
			1st n.n.	2nd n.n.	3rd n.n.
1	10	12	4	4	0
2	30	40	4	4	12
3	88	140	24	0	12
4	184	308	36	0	24
5	338	584	52	4	36
6	550	972	64	4	60
7	848	1528	96	0	72
8	1232	2248	120	0	96
9	1722	3172	148	4	120
10	2318	4304	172	4	156
11	3048	5700	216	0	180

[a] n.n. = nearest neighbor.

TABLE 1.2

Bond Energies[a]

Si—Si	42.2
Si—H	70.4
Hybridized bonds	$(E_b = \frac{1}{2} k x^2)$
First nearest neighbor	42.2
Second nearest neighbor	112.5
Third nearest neighbor	267.3

[a] kcal/mol.

TABLE 1.3

Total Cluster Bond Energy

Cluster size (no. of shells)	Si cluster (kcal/mol)	Si:H cluster (kcal/mol)
1	112.5	163.3
2	<u>183.8</u>	<u>150.1</u>
3	114.2	123.9
4	113.7	116.6
5	109.2	111.2
6	<u>109.5</u>	<u>107.4</u>
7	103.5	103.9
8	101.9	101.7
9	100.2	100.0
10	99.7	98.5
11	97.7	97.2

shows maximum energies for stable two- and six-layered Si clusters (as noted by italic numbers) and for no other size. Further, calculations show that these clusters have minimum lattice energies with the lowest for the six-layered cluster. The latter calculation involves treating the second and the fifth and sixth layers in tension for the two- and six-layered clusters, respectively. In addition, note that less bond energy is required for the formation of correspondingly sized hydrogenated clusters; however, by themselves the latter clusters are not stable without hybridized Si bonds to maintain surface tension and without intercluster pressure. With reference to Table 1.1, the number of third nearest-neighbor hybridized bonds reaches a maximum for two- and six-layered clusters. This is a necessary bridge to tie parallel $\langle 110 \rangle$ chains of Si atoms together for surface tension, as shown in Fig. 1.4.

2. Cluster Hierarchy

A cluster up to this point is a stable or metastable entity that may exist in all phases of condensed matter but is limited in size by thermal fluctuations and the time needed to form bonds, as discussed for the quenching process in Section 1.3.

A. CRYSTALLINE AND POLYCRYSTALLINE STATES

Under near-equilibrium conditions, only the six-layered cluster can form at the solid substrate or the film layer interface in contact with a liquid or vapor phase. The facets on the six-layered cluster are well-defined {110} planes, and hence will perfectly stack together. However, in the solid crystalline state, a seventh layer of Si must diffuse between the clusters to bond them together. If the clusters do not pack together perfectly over long distances, the resultant structure will be polycrystalline, and in some instances when the clusters pack together but are not properly oriented twin boundaries will be formed.

There are many physical measurements of Si single crystals that can be interpreted in terms of six-layered clusters, e.g., a 7×7 LEED pattern for a (111) surface [58] is indicative of a macroscopic array of six-layered clusters bonded together by a seventh layer. In particular, the seventh layer between the clusters is an internal surface that may host Si and/or substitutional atoms and vacancies, and it is also the boundary for slip planes. This layer is of major importance when the crystalline state melts. The heat of formation is the difference in bond formation energy between the seven-layered cluster and the three-layered cluster, or 11 kcal/mol (Table 1.3). This result can be obtained by several transformation routes between two- and six-layered stable cluster size changes with intercluster bonding by the third and seventh layers of atoms.

B. AMORPHOUS STATE AND LIQUID PHASE

For a high density of Si atoms or SiH_4 molecules in the liquid or vapor phase that are rapidly quenched by deposition onto a cold substrate surface at a rate that limits bond formation, a metastable two-layered cluster can be formed that is ~ 8.5 Å in diameter (Table 1.4). Unique to the two-layered cluster is that the surface appears more like a sphere of atoms. The {110} surface planes are barely distinguishable as facets. In addition, a cluster with a third layer of atoms, except for four atoms at the $\langle 111 \rangle$ corners, is an octahedron bounded by {111} facets and will not close pack. Therefore, clusters with random orientations at the substrate surface will pack more like spheres and be weakly bonded together by atoms in the third layer.

Evidence for this formation is substantiated by several structural and thermal observations:

TABLE 1.4

Density of Clusters

Cluster size (no. of shells)	Density (no./cm^3)	Mean diameter (Å)	fcc Spacing (Å)
1	3.12×10^{21}	4.2	3.8
2	9.25×10^{20}	8.5	7.7
3	3.90×10^{20}	12.7	11.5
4	2.00×10^{20}	17.0	15.4
5	1.16×10^{20}	21.2	19.2
6	7.28×10^{19}	25.4	23.0
7	4.88×10^{19}	29.7	26.9

(1) The short-range order determined by the mean number of first, second, etc., nearest neighbors in a two-layered cluster (Table 1.5) shows equal numbers of second and third nearest neighbors, and one broad second peak is obtained and observed [16].

(2) Hydrogenated Si bonds are at the surface of the cluster and appear like $(SiH_x)_n$ molecules, where $x = 1$, 2, and 3 and $n = 1$ and 2 (Table 1.6). These molecules are observed [30], and their relative abundance is determined by the substrate deposition temperature and rf power, which varies the quench rate for forming only two-layered clusters or clusters weakly bonded by a third layer.

(3) Calculated densities of hydrogenated amorphous thin films of Si agree closely with the atomic percent H in the films [30, 59] (Table 1.7). In addition, Table 1.8 shows that the two-layered cluster can accommodate the highest percentage of H—Si bonds, 44.4%, and up to an additional 13% H can occupy the third-layer sites not bonded directly to Si in the second shell. This calculation assumes that the third layer exists but that it is not uniformly shared between neighboring clusters. Films with more than

TABLE 1.5

Number of Nearest Neighbors[a]

Cluster size (no. of shells)	1st n.n.	2nd n.n.	3rd n.n.	4th n.n.
1	2.40	3.60	2.40	0.60
2	2.67	5.60	4.80	1.80
3	3.18	7.50	6.55	2.66
⋮				
∞	4	12	12	6

[a] n.n. = nearest neighbor.

TABLE 1.6

$(SiH_x)_n$ Distribution[a]

Cluster size (no. of shells)	SiH_3	SiH_2	$(SiH_2)_2$	SiH
1	0	6	0	4
2	4	0	6	4
3	4	6	0	48
4	0	12	6	72
5	0	30	0	124
6	4	24	6	172
7	4	30	0	264
8	0	36	6	336

[a] Number of molecules.

TABLE 1.7

Amorphous Si:H *Alloy Density* [a]

Atom percent H	Crystalline density	Amorphous density (g/cm^3) [b]	Experimental density	Reference
0	2.331	2.25	2.25	[59]
5	2.219	2.13		
10	2.106	2.02		
15	1.994	1.91		
20	1.882	1.81		
25	1.769	1.70	1.70	[59]
30	1.657	1.59		
35	1.544	1.48	1.47	[59]
40	1.432	1.37		
44.4	1.332	1.28		

[a] Two-layered cluster with H and Si in the third layer.
[b] The amorphous density is $\sim 96\%$ of the crystalline density.

TABLE 1.8

Hydrogen Content in Amorphous Si:H *Alloys*

Cluster Size (no. of shells)	No. of Si	No. of H	Atom percent H	Atom percent H_{max}
1	10	6	37.5	61.5
2	30	24	44.4	57.1
3	88	40	31.2	45.0
4	184	66	26.4	39.5

TABLE 1.9

Si *Heats of Formation*

Transformation	Cluster model (kcal/mol)	Experiment (kcal/mol)
Amorphous Si:H to polycrystalline Si	40, 46	44, 50 [59]
Crystal Si to liquid Si	11	12
Liquid Si to vapor Si	107	106

$\sim 35\%$ H are like "butter" and are not stable under atmospheric conditions [60].

(4) Annealing hydrogenated Si films above 350°C drives off the H, and above $\sim 500°$C the polycrystalline state forms [59]. When the H bonds break, stable six-layered clusters form, and then a seventh layer of atoms bonds the clusters together. The cluster formation is relatively spontaneous, while the bonding layer is principally formed by diffusion. The two

heats of formation are ∼ 40 and 46 kcal/mol, respectively, and are in close agreement with experiment [59] (Table 1.9).

The solid-to-liquid phase transition is almost the reverse of crystallizing the amorphous phase. The seventh layer of atoms bonding the clusters together diffuses away, and the six-layered clusters break down into two-layered clusters. Similarly, the liquid-to-vapor phase transition occurs by the dissociation of two-layered clusters into small chainlike clusters or molecules. Their respective heats of formation are summarized in Table 1.9.

The cluster model suggests that the heats of formation at any transformation only involve the bond changes from one cluster size to another and/or the breaking or forming of bonds between clusters. The existence of clusters in a solid phase means that there is an internal network of surfaces. This supports Phillips' model that proposes internal surfaces, edges, and corners to explain electron transport in amorphous Si:H alloy films [26].

1.6   CLUSTER MODEL APPLIED TO III–V COMPOUNDS

Applying the cluster model to III–V compounds gives stable and metastable cluster sizes one shell larger than for Si [43]. The basic reasons for the difference are that hybridized second nearest-neighbor bonds cannot form and only the three- and seven-layered clusters and, in addition, the fourth and eighth layers are stoichiometric. Cohesive lattice energy calculations also show that the seven-layered cluster is the most stable [43]. Therefore, the ability to form an amorphous III–V compound should be more difficult because the three-layered cluster is less spherical and the group V elements are highly volatile. The resultant material will probably be a disordered amorphous state, because the group V elements readily quench into an amorphous state [27].

The cluster model still requires further study in theory and in experimental trials for verification. However, the consistency of the model appears to interrelate the hierarchy in the condensed states of solids and other phases of matter.

ACKNOWLEDGMENTS

I wish to thank T. W. Barbee, Jr., A. E. Blakeslee, M. H. Brodsky, C. H. L. Goodman, J. C. Knights, S. C. Moss, and M. P. Shaw for their help in acquainting me with the amorphous and polycrystalline states and the relevant literature.

REFERENCES

1. J. J. Gilman (ed.), "The Art and Science of Growing Crystals." Wiley, New York, 1963.
2. B. Chalmers, "Principles of Solidification." Wiley, New York, 1964.
3. J. C. Brice, "The Growth of Crystals from the Melt." North-Holland Publ., Amsterdam, 1965.
4. K. A. Jackson, D. R. Uhlmann, and J. D. Hunt, *J. Cryst. Growth* **1**, 1 (1967).
5. B. R. Pamplin (ed.), "Crystal Growth." Pergamon, Oxford, 1976.
6. F. F. Abraham, "Homogeneous Nucleation Theory." Academic Press, New York, 1974.
7. F. L. Binsbergen, *Prog. Solid State Chem.* **8**, 189 (1973).
8. D. W. Pashley, *Adv. Phys.* **14**, 327 (1965).
9. H. Sato, *Ann. Rev. Mater. Sci.* **2**, 217 (1972).
10. B. A. Joyce, *Rep. Prog. Phys.* **37**, 363 (1974).
11. K. L. Chopra, "Thin Film Phenomena," Chapter IV. McGraw-Hill, New York, 1969.
12. C. A. Neugebauer, "Handbook of Thin Film Technology" (L. I. Maissel and R. Glang, eds.), Chapter 8. McGraw-Hill, New York, 1970.
13. S. Takagama, *J. Mater. Sci.* **11**, 164 (1976).
14. T. N. Barbee, Jr., W. H. Holmes, D. L. Keith, and M. K. Pyzna, *Thin Solid Films* **45**, 591 (1977).
15. R. Grigorovici and R. Manaila, *Thin Solid Films* **1**, 343 (1967).
16. M. H. Brodsky, R. S. Title, K. Weiser, and G. D. Pettit, *Phys. Rev. B* **1**, 2632 (1970).
17. D. E. Polk, *J. Non-Cryst. Solids* **5**, 365 (1971).
18. D. E. Polk, *Acta Metall.* **20**, 485 (1972).
19. N. J. Shevchik, *Phys. Rev. Lett.* **31**, 1245 (1973).
20. W. Paul, G. A. N. Connell, and R. J. Temkin, *Adv. Phys.* **22**, 529 (1973).
21. R. J. Temkin, W. Paul, and G. A. N. Connell, *Adv. Phys.* **22**, 581 (1973).
22. J. D. Joannopoulos and M. L. Cohen, *Phys. Rev. B* **7**, 2644 (1973).
23. S. C. Moss, *Ann. Rev. Mater. Sci.* **3**, 293 (1973).
24. S. C. Moss, *Proc. Int. Conf. Amorphous Liquid Semicond.*, *5th Garmisch-Partenkirchen* p. 17 (1973).
25. P. H. Gaskell, *J. Non-Cryst. Solids* **32**, 207 (1979).
26. J. C. Phillips, *Phys. Rev. Lett.* **42**, 1151 (1979).
27. Y. Kato, T. Shimada, Y. Shiraki, and K. F. Komatsubara, *J. Appl. Phys.* **45**, 1044 (1974).
28. D. K. Paul and S. S. Mitra, *J. Non-Cryst. Solids* **18**, 407 (1975).
29. T. Shimada, Y. Kato, Y. Shiraki, and K. F. Komatsubara, *J. Phys. Chem. Solids* **37**, 305 (1976).
30. J. C. Knights, *Jpn. J. Appl. Phys.* **18**, 101 (1978).
31. J. C. Knights and R. A. Lujan, *J. Appl. Phys.* **49**, 1291 (1978).
32. M. H. Brodsky, *Thin Solid Films* **50**, 57 (1978).
33. G. A. N. Connell, R. J. Temkin, and W. Paul, *Adv. Phys.* **22**, 643 (1973).
34. K. L. Narasimhan and S. Guha, *J. Non-Cryst. Solids* **16**, 143 (1974).
35. G. A. N. Connell and J. R. Pawlik, *Phys. Rev. B* **13**, 787 (1976).
36. M. H. Brodsky, M. Cardona, and J. J. Cuomo, *Phys. Rev. B* **16**, 3556 (1977).
37. A. Barna, P. B. Barna, G. Radnoczi, L. Toth, and P. Thomas, *Phys. Status Solidi (a)* **41**, 81 (1977).
38. D. W. Bullett and J. J. Kelly, *J. Non-Cryst. Solids* **32**, 225 (1979).
39. D. Bermejo and M. Cardona, *J. Non-Cryst. Solids* **32**, 405 and 421 (1979).
40. R. Wang and M. D. Merz, *Nature (London)* **257**, 370 (1975).
41. C. H. L. Goodman, *Nature (London)* **257**, 370 (1975).
42. J. J. Gilman, *Phil. Mag. B* **37**, 577 (1978).

43. B. L. Mattes, *J. Vac. Sci. Technol.* **13**, 816 (1976).
44. T. L. Chu, H. C. Mollenkoph, and S. S. Chu, *J. Electrochem. Soc.* **122**, 1681 (1975).
45. T. L. Chu, *J. Cryst. Growth* **39**, 45 (1977).
46. B. L. Mattes and R. K. Route, *J. Crystal Growth* **27**, 133 (1974) or B. L. Mattes, *CRC Crit. Rev. Solid State Sci.* **5**, 457 (1975).
47. F. C. Frank, *Proc. R. Soc. London Ser. A* **215**, 43 (1952).
48. N. F. Mott, *Proc. R. Soc. London Ser. A* **215**, 1 (1952).
49. H. Reiss, *J. Appl. Phys.* **39**, 5045 (1968).
50. K. Heinemann and H. Poppa, *Thin Solid Films* **33**, 237 (1976).
51. A. Glasner, S. Skurnik-Sarig, and M. Zidon, *Israel J. Chem.* **7**, 649 (1969).
52. A. Glasner and M. Zidon, *J. Cryst. Growth* **21**, 294 (1974).
53. J. Farges, *J. Cryst. Growth* **31**, 79 (1975).
54. T. Takagi, I. Yamada, and A. Sasaki, *J. Vac. Sci. Technol.* **12**, 1128 (1975).
55. J. B. Theeten, R. Madar, A Mircea-Roussel, A. Rocher, and G. Laurence, *J. Cryst. Growth* **37**, 317 (1977).
56. P. L. Vuillermoz, A. Laugier, and P. Pinard, *Phys. Status Solidi (b)* **63**, 271 (1974).
57. H. Mell and M. H. Brodsky, *Thin Solid Films* **46**, 299 (1977).
58. B. L. Mattes, *J. Vac. Sci. Technol.* **13**, 360 (1976).
59. M. H. Brodsky, M. A. Frisch, and J. F. Ziegler, *Appl. Phys. Lett.* **30**, 561 (1977).
60. J. C. Knights, private communication.

2 | Electrical and Optical Properties of Amorphous Thin Films

DAVID EMIN

Sandia National Laboratories
Albuquerque, New Mexico

2.1 INTRODUCTION

During the past decade considerable efforts have been expended in attempts to characterize and understand the electronic properties of amor-

POLYCRYSTALLINE AND AMORPHOUS
THIN FILMS AND DEVICES

phous semiconductors. Through the course of these endeavors there have been radical changes in the concepts applied to these materials. For example, basic tenets such as the undopability of amorphous semiconductors have been modified or abandoned, while new ideas such as the significance of polaron effects have been promoted to prominence. These changes have accompanied an alteration in materials preparation techniques, an expansion of available data, a widening of phenomena studied, and a utilization of new experimental probes. Presently the field is still in a state of flux, with a number of issues yet unresolved. The intent of this chapter is (1) to present the original ideas, (2) to describe how and why they have been changed or questioned, and (3) to indicate those issues which are currently objects of concern. It will concentrate on presenting the physical principles which underlie various models and emphasize those features which distinguish different views.

The chapter begins with a brief discussion of the structure of covalently bonded amorphous semiconductors. Following this, differing definitions of electronic localization are presented and various models of localization in amorphous semiconductors are described. A discussion of the dark transport associated with various localization models then follows. Attention is then shifted to the properties of amorphous semiconductors during and after illumination, and the exposition addresses the nature of the absorption associated with the different models. The photon-induced electronic properties of these systems are then considered. The chapter concludes with a succinct synopsis of the state of current knowledge in which some of the outstanding issues are enumerated.

2.2 STRUCTURE

To define the structure of a solid, the equilibrium positions of all the atoms which constitute the material must be enumerated. In the case of a crystal, the translational degeneracy of the regular lattice reduces this task to that of specifying a manageable number of position vectors. However, this crucial simplification is no longer feasible in the instance of an amorphous semiconductor. Thus the structure of a given amorphous solid cannot be defined with the precision that characterizes perfect crystals. However, despite the severely limited knowledge of amorphous structures compared with those of crystals, some very general notions have been developed.

It has been recognized that the structure of a covalently bonded amorphous semiconductor is topologically distinct from that of a crystalline counterpart. That is, the transformation between crystalline and amorphous structures involves more than the stretching and bending of bonds. Rather, it requires the breaking and reforming of bonds. Furthermore, since the energy associated with the bonding of a covalently bonded semiconductor is both large and strongly dependent on bond lengths and angles, it is often stated that, while there is no long-range order in a covalently bonded amorphous semiconductor, there is nonetheless short-range order. In this view, the variations of the bond lengths and bond angles involved in connecting one atom to its neighbors are fractionally very small.

Within such a picture, one can construct model amorphous networks by establishing algorithms which define the conditions under which an atom will occupy a given location and bond with its neighbors. For example, ideal amorphous networks can be created by demanding that every atom fulfill its (nominally) appropriate bonding requirements with the associated bond lengths and bond angles lying within prescribed tolerances. One can then characterize the resulting structures by their density, radial distribution function (rdf), and ring statistics. The rdf describes the distribution of interatomic separations associated with the structure. The ring statistics are associated with viewing the material as being made up of a distribution of rings composed of the bonds as elements. Specifically, enumerating the distribution of irreducible rings (those which do not enclose other rings) as a function of the number of ring elements constitutes designating the ring statistics of a structure. For example, crystalline silicon is composed of only six-membered rings, while some models of amorphous silicon (a-Si) yield large fractions of odd-membered rings (five- and seven-membered rings) [1]. Modeling schemes can also be carried out with bonding rules that permit the existence of unfulfilled ("dangling") bonds. Indeed, one can also envision situations (such as with chalcogenide atoms) in which some electrons that are usually nonbonding may be used for bonding [2, 3]. In these instances the densities of such "defects" are additional characterizing parameters of the structures.

A prevailing concern about the morphology of amorphous semiconductors is that the surface regions may differ qualitatively from the bulk. For example, relatively large densities of dangling bonds may be attributed to surface regions, although in some circumstances surface reconstruction may eliminate a fraction (perhaps large) of such dangling bonds. Nonetheless, as a result of structural differences, the electronic defect states at a surface can differ from those in the bulk. Thus defect-related properties may be a function of film thickness as the ratio of surface to bulk defect states is altered. Indeed, a conductive surface layer has been produced in a bulk chalcogenide glass with sufficient annealing [4]. Furthermore, surfaces can be associated with internal voids which develop as the amorphous network is formed, or as clusters of a material coalesce to form larger structural units. Thus considerations of surface states pertain to internal as well as external surfaces.

About a decade ago it was often assumed that most frequently studied amorphous semiconductors could be approximated by networks devoid of dangling bonds [5, 6]. Similarly, it was presumed that the network would readily accommodate atoms of various valences, thereby precluding the introduction of donors or acceptors into the material. However, experiments on tetrahedrally coordinated amorphous films (a-Si and a-Ge) have indicated the presence of large densities of dangling bonds associated with defect states deep within the semiconductor gap [7]. Furthermore, with the elimination (pacification) of these dangling bonds (e.g., by introducing hydrogen) the defect density can be reduced sufficiently so that n- and p-type doping is readily observable [8]. However, even in these cases most of the dopant atoms appear to be incorporated into the network without producing donors or acceptors. In chalcogenide glasses, on the other hand, dangling bonds are not typically observed, and doping has not been reported. Presumably this is because these glasses are softer and have a lower coordination than the stiffer tetrahedrally coordinated materials.

2.3 LOCALIZATION

1. Definitions

The question of the localization of electronic charges is central to consideration of the electronic properties of amorphous semiconductors. It is therefore appropriate to begin a discussion of localization by formally defining the notion as follows. A dichotomy exists between those eigenfunctions of an electron in a static potential which are normalizable and those which are not. If the wave function associated with a given eigenstate is normalizable within an *infinite* volume, the eigenstate is said to be localized. Otherwise it is termed extended.

To ascertain whether particular electronic states are localized or extended, the system must be probed in search of a response characteristic of one or the other situation. Frequently one employs a transport measurement. For instance one may attempt to determine the carrier's intrinsic drift mobility, its response to an electric field. If the mobility is high ($\gg 1$ cm^2/V s) and decreases with increasing temperature, the measurements are typically viewed in terms of a scattering picture in which the zeroth-order states are extended. Alternatively, if the mobility is low ($\ll 1$ cm^2/V s) and increases with increasing temperature, the transport is usually interpreted within a hopping picture in which the zeroth-order states are localized. Thus these measurements may be viewed as providing a transport-based operational definition of localized and extended states. However, the formal and operational definitions are not always equivalent [9]. Discrepancies can arise when the electronic eigenstates of the static system are not the appropriate zeroth-order states for the phenomena being observed in probing the system. This situation occurs in transport when the uncertainty of the electronic energy associated with interactions with the atomic vibrations becomes larger than a characteristic electronic energy. For example, in the case of conduction in a very narrow band of a crystal, if the energy uncertainty due to scattering exceeds the bandwidth, the appropriate zeroth-order states for transport are localized (Wannier) states and the motion is described as hopping. Thus even in a crystal, where the electronic eigenstates are extended (in terms of the formal defintion), transport can be of the hopping variety, implying (with the operational definition) that the involved states are localized. In discussing amorphous semiconductors, as elsewhere in solid-state physics, the adjectives "localized" and "extended" are used in reference to both electronic eigenstates and appropriate zeroth-order states. The reader must judge from the context which is the relevant definition.

2. Starting Equations

A convenient starting point for discussing localization phenomena is the set of matrix eigenvalue equations for a carrier in a (generally disordered) solid in which the electronic basis states are taken to be local (Wannier-type) functions. Specifically, employing the adiabatic principle, one writes

$$\left(E_i - H_{\mathrm{vib}} - \epsilon_{\mathbf{g}}\right)a_{\mathbf{g}}^i = \sum_{\mathbf{g}' \neq \mathbf{g}} J_{\mathbf{g},\,\mathbf{g}'} a_{\mathbf{g}'}^i \qquad (2.1)$$

where E_i is the energy of the ith eigenstate of the system, H_{vib} the Hamiltonian which describes the atomic vibrations of a carrier-free system, $\epsilon_{\mathbf{g}}$ the electronic energy associated with the carrier occupying the site labeled by the position vector $\mathbf{g}$, $J_{\mathbf{g},\,\mathbf{g}'}$ the transfer integral linking site $\mathbf{g}$ with site $\mathbf{g}'$, and $a_{\mathbf{g}}^i$ and $a_{\mathbf{g}'}^i$ the amplitudes of the ith eigenfunction associated with the carrier occupying sites $\mathbf{g}$ and $\mathbf{g}'$. With the exception of the energy eigenvalue E_i, these quantities are generally functions of the atomic displacements.

3. Anderson Localization

If, however, one ignores this dependence of the electronic energies and transfer integrals on atomic displacements (neglect of the electron–lattice interaction), the equations reduce to the electronic tight-binding result:

$$(\epsilon_k - \epsilon_{\mathbf{g}})b_{\mathbf{g}}^k = \sum_{\mathbf{g}' \neq \mathbf{g}} J_{\mathbf{g}, \mathbf{g}'} b_{\mathbf{g}'}^k \qquad (2.2)$$

where ϵ_k is the eigenvalue of the kth eigenstate of the purely electronic system and $b_{\mathbf{g}}^k$ the occupation amplitude of site $\mathbf{g}$ associated with the kth electronic eigenfunction; each such amplitude is related to the respective coefficient of the more general equation, Eq. (2.1), by the simple proportionality $a_{\mathbf{g}}^i = b_{\mathbf{g}}^k x^n$, where x^n is the nth eigenfunction of the atomic vibrations of the carrier-free solid. Disorder appears in these equations through the positions vectors, which refer to an amorphous arrangement of sites, and via the local electronic energies and transfer integrals, which possess a distribution of values consistent with the noncrystalline structure. To facilitate solution of these equations, additional simplifications are often introduced. The disordered solid has been modeled as a regular array of sites in which the only nonzero transfer integrals are those linking directly bonded (nearest-neighbor) sites. Furthermore, these finite transfer integrals have often been assigned a common value J. Thus the sole property which manifests the pressure of disorder, in this model, is that the local electronic energies are assumed to possess a random distribution of values. This is the Anderson model [10]. Localization within this model has been termed Anderson localization.

The phenomenon of Anderson localization can be understood by noting that a particle moves through a material via a succession of real (energy-conserving) transitions which transport it between different locales. By imposing a distribution of local-site energies, the ability to find extended paths which link (nearly) degenerate sites is impaired. In this situation, states confined to a restricted region of the material, Anderson-localized states, may be formed. In particular, a transition is regarded as real (energy-conserving), linking (nearly) degenerate states, if the energy uncertainty associated with the transition between the two states is greater than the magnitude of the difference between the energies of the two states. For example, a direct transition between adjacent sites $\mathbf{g}$ and $\mathbf{g}'$ is real if $|J_{\mathbf{g}, \mathbf{g}'}| > |\epsilon_{\mathbf{g}} - \epsilon_{\mathbf{g}'}|$. Those transitions which are not real are termed virtual; they involve a modification of the local states without (directly) permitting the escape of a particle. For example, virtual transitions from a site characterize the expansion of a local wave function to encompass a finite number of surrounding sites. These expanded states form the basis set between which the real transitions associated with extended motion can occur. With regard to the possibility of real transitions between these basis states, one notes that the likelihood of finding two states within a given energy interval increases as one considers pairs of basis states with increasingly large separations. However, with increasing separation the effective transfer energies associated with these pairs tend to decrease. Thus, if the transfer energy falls off sufficiently fast with separation, real transitions between well-separated pairs are rare occurrences and localized states can be formed. Anderson argued that, with a sufficiently broad spread of local-site energies, $|\Delta\epsilon_g| \gg |J|$, all the states of the band will be localized. With a less drastic distribution of site energies, those states of lowest density which reside near the edges of the bands (in the band tails) tend to be localized. It has been inferred that there are distinct energies which serve as demarcations between extended and localized states. These are termed *mobility edges* [5, 6]. Within this scheme the separation between the upper valence-band mobility edge and the lower conduction-band mobility edge is designated the *mobility gap*.

The effect of including disorder of the intersite transfer energies, the $J_{g,g}$'s, can be discussed within the framework of Anderson localization. States associated with exceptionally large transfer energies will tend to be shifted far from the band center into the band tails, where the low density of states facilitates localization. Sites with exceptionally small transfer energies will tend to generate states near the middle of the band among a high density of states of similar energy. In analogy with the original Anderson localization problem, the question of localization of states lying increasingly deep within the band hinges on the competition between the increase in the density of states and the decrease in the energy intervals associated with real transitions.

In summary, amorphous semiconductors have been simulated by model lattices possessing a spread of local-site energies and, sometimes, a distribution of transfer energies. While the role of topological disorder in localization has not been investigated, it nonetheless appears that the Anderson mechanism would not be qualitatively altered in passing from models of deformed crystals to those of (topologically distinct) amorphous structures. Typically, the disorder energies envisioned in amorphous materials (associated, say, with bond angle fluctuations) lead, within the Anderson model, to a picture of the band structure of an amorphous semiconductor, which, as illustrated in Fig. 2.1, possesses rather narrow ($\sim 10^{-1}$ eV) bands of localized states adjacent to the mobility edge within the mobility gap. Thus, within this scheme, amorphous semiconductors only depart from crystalline semiconductors in that they contain a number of localized traplike states within the mobility gap [5, 6]. Their electronic properties have therefore been expected to bear a close resemblance to those of common high-mobility crystalline semiconductors [5, 6]. That is, the charge carriers which exist above the mobility edge are presumed to behave much like carriers in crystalline silicon, manifesting high intrinsic

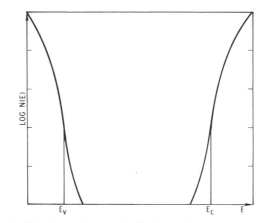

Fig. 2.1 The density of states for one idealized view of an amorphous semiconductor. E_c and E_v denote the mobility edges associated with the conduction and valence bands, respectively.

mobilities, $\mu \gg 1$ cm^2/V s, and interacting only weakly with the atomic vibrations. In actuality, while the electrons in a-SiO$_2$ manifest such high mobilities [11], for most well-studied amorphous semiconductors (such as the chalcogenide glasses, amorphous silicon, germanium, and arsenic, and the transition-metal-oxide glasses) there is little direct evidence supporting this hypothesis. Rather, as will be discussed subsequently in this chapter, many experiments suggest that a far different situation prevails: The carriers possess very low mobilities and interact strongly with the atomic vibrations, forming small polarons. Therefore we shall now address the question of small-polaron formation.

4. Small-Polaron Formation

Within the tight-binding scheme of Eq. (2.1), the interactions of an electronic carrier with the atomic displacements of a solid are contained in the dependence of the ϵ_g's and the $J_{g,g'}$'s on atomic displacements. It is these interactions which will now be discussed. To begin, consider a stationary electron which is introduced at one of the many geometrically equivalent sites of a perfect crystal. Such a carrier will generally exert forces on the atoms surrounding it, which produces displacement of the equilibrium positions of these atoms. These displacements create a potential well for the carrier at the expense of introducing a strain in the material. If the induced potential well is appropriate in depth and spatial extent, the carrier may occupy a bound state within it. In this instance the carrier is said to be *self-trapped*; the carrier cannot move without an alteration of the positions of the atoms surrounding it. The term *polaron* refers to the quasi-particle unit comprising the self-trapped carrier and the associated atomic displacement pattern. If the spatial extent of the carrier's wave function is severely localized on the scale of interatomic distances, the polaron is referred to as being small. If the wave function is of much larger extent, it is said to be a large polaron.

An important aspect of polaron formation in solids is that the size of the polaron that can be formed depends upon the range of the electron–lattice interaction. In particular, in solids generally (and especially in covalent solids) a significant component of the electron–lattice interaction is short ranged; it is associated with the interactions of a carrier with the atoms immediately adjacent to it. In such situations there are states of two distinct types for a carrier in a deformable medium: Either the carrier spreads out over very many sites, or it shrinks to the smallest size that is commensurate with the particular lattice [12–15]. In other words, either the polaron effects are relatively minor or they are of major importance with the carrier shrinking to form a small polaron. To understand this, consider the ground-state energy $E(R)$ of an electron in a deformable continuum as a function of the radius of the electron's wave function R [15]. As illustrated in Fig. 2.2, for a short-range interaction the lowering of the potential energy of the electron increases with decreasing R as R^{-3}, while the self-induced potential well deepens. At the same time the kinetic

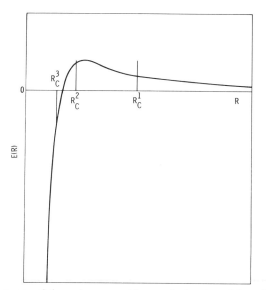

Fig. 2.2 The energy of the system comprising an electron in a deformable medium as a function of R, a parameter related to the spatial extent of the charge's wave function, within the adiabatic approximation. R_C^1, R_C^2, and R_C^3 depict three possible cutoff values for R.

energy associated with compressing the electronic wave function increases with decreasing R as R^{-2}. The net effect is to allow only two stable solutions: a solution at infinite R corresponding to a spreadout electronic state with little lattice deformation, and a solution at $R = 0$ corresponding to small-polaron formation. In a real material the magnitude of the lowering of the potential energy saturates at a maximum value when R is less than a minimum value R_c of the order of the interatomic separation. Thus the curves in Fig. 2.2 only have meaning for $R \geqslant R_c$. The three possible real situations are shown in Fig. 2.2: (1) R_c is such that there are two solutions, the small-polaron solution being the stable one; (2) R_c is such that, while the small-polaron solution exists, it is only metastable; and (3) only the weak-coupling (large-R) solution exists.

As indicated in Fig. 2.2, when the small-polaron state exists, an energy barrier also exists which separates it from the nonpolaronic solution. Surmounting or tunneling through this barrier corresponds to the atoms classically or quantum mechanically undergoing rearrangements that correspond to the change of state. The height of this barrier tends to diminish with an increase in either the carrier's effective mass or the strength of the electron–lattice interaction. Furthermore, if the particle experiences an additional potential (associated with defects or disorder) which tends to confine it, the barrier may be reduced or eliminated. This is illustrated in Fig. 2.3a and b, respectively, where the results for an electron attracted by a coulombic center are shown; the coulombic center contributes an attractive component to the electron's potential, which varies as R^{-1} [15]. Furthermore, the presence of a significant barrier manifests itself in a time delay for small-polaron formation [16, 17]. For an injected carrier to form a small polaron, the atoms of the material must assume an appropriate displacement pattern. In the absence of a barrier, the requisite time is of the order of a vibrational period. However, when a barrier is present, the time interval can be very much longer. In such instances the delay time associated with atomic rearrangements decreases with increasing temperature. While there is limited information about the delay time for small-polaron formation in amorphous semiconductors, the available evidence suggests that it is near the (minimal) time of a picosecond in a-SiO$_2$ [18] and chalcogenide glasses [19]. As will be discussed later in this chapter, this rapid small-polaron formation has been associated with significant localization in the precursor nonpolaronic state.

So far this discussion of polaronic localization has been concerned with a carrier that adiabatically adjusts to the deformable continuum in which it is contained. Additional insight into the problem of polaronic localization can be gained by considering small-polaron formation in a discrete system in which the requirement of adiabaticity is not imposed. First, consider the

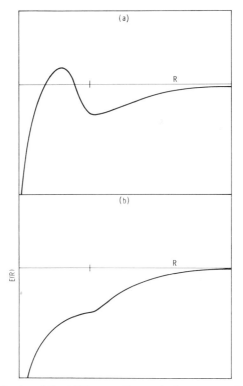

Fig. 2.3 Modification of the $E(R)$ versus R curve in the presence of a coulombic well is illustrated for two different values of the physical parameters. (See Emin and Holstein [15].)

motion of a small polaron between adjacent geometrically equivalent sites in a lattice. As illustrated in Fig. 2.4, the motion of the particle is associated with an alteration of the positions of the atoms of the solid [20]. Concomitantly, the complete intersite transfer integral is the product of an electronic transfer integral and a vibrational overlap factor associated with the changed atomic positions. If the carrier-induced atomic displacements are much larger than the amplitudes of the zero-point atomic motion, the

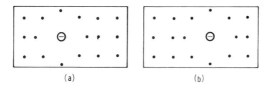

Fig. 2.4 The atomic displacement pattern about a self-trapped electron is illustrated for the charge occupying each of two adjacent sites: situations (a) and (b).

vibrational overlap factor is much less than unity—typically orders of magnitude less than 10^{-2}. Thus the intersite small-polaron transfer matrix element is greatly reduced relative to that for nonpolaronic transfer. In other words, the small-polaron bandwidth, which is proportional to the transfer integral, is orders of magnitude smaller than that for nonpolaronic carriers. As a result, even rather modest disorder will produce Anderson localization in a system in which the carriers form small polarons. Furthermore, even in the absence of disorder, the energy uncertainty associated with the scattering of a small polaron can be larger than the small-polaron bandwidth. In such instances, which are equivalent to the mean free path being smaller than the intersite separation, the transport is best characterized as being phonon-assisted hopping motion. Thus, in the transport sense, small polarons can be localized even in a perfect crystal. Indeed, small-polaron hopping has been observed in a wide variety of crystalline semiconductors [9].

Study of the adiabatic eigenstates of an electron in a three-dimensional continuum has established that, with a short-range interaction, only small-polaronic and nonpolaronic (but not intermediate polaronic) eigenstates can exist [13–15]. With the presumption, supported by this study, that no intermediate polaron can exist in a system with a short-range interaction, one can construct a variational approach to the eigenstates of an electron in a deformable simple-cubic lattice which transcends the assumption of adiabaticity [13, 14]. This type of study enables one to investigate the conditions under which nonpolaronic and small-polaronic eigenstates occur (are self-consistent). Typical results of such an investigation are shown in Fig. 2.5, where the energy spectrum of the eigenstates is plotted as a function of the electron–lattice coupling strength (measured in units of the ratio of the small-polaron binding energy to the optical phonon energy). The energies plotted on the ordinate are measured in units of the half-width of the (nonpolaronic) conduction band, $6J$. As shown, for $6J/h\omega_0 = 10$, the two separate classes of solution coexist within a range of coupling strengths, $C_{small} \leqslant E_b/h\omega_0 \leqslant C_{weak}$. In the adiabatic limit, $6J/h\omega_0 \to \infty$, both the small-polaron and conduction-band states coexist for all values of $E_b/h\omega_0$, in accord with the continuum study. In the narrow-band limit, $6J/h\omega_0 \ll 1$, only the small-polaron solution appears.

The conditions which characterize the existence of each of the two solutions can be described physically. The small-polaronic solution becomes dynamically unstable (for $E_b/h\omega_0 \leqslant C_{small}$) when the energy of the small-polaron system decreases as a result of a reduction in the distortion about the carrier. Specifically, in such an instance with a decrease in the deformation, the energy lowering due to intersite transfer increases (as a result of enhanced vibrational overlap) more rapidly than the local single-

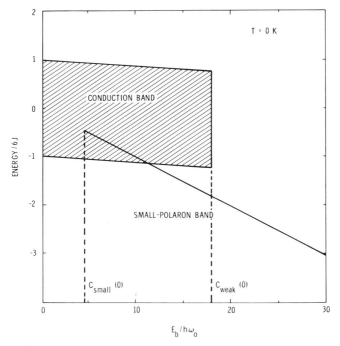

Fig. 2.5 The energy spectrum of an electron in a deformable lattice, in units of $6J$, is plotted against the electron–lattice coupling strength $E_b/\hbar\omega_0$. The temperature is taken to be absolute zero. (See Emin [14].)

site energy increases. The nonpolaronic conduction-band solution becomes dynamically unstable (for $E_b/h\omega_0 \geqslant C_{weak}$) when the time of stay of the electron at any given site exceeds that associated with the creation of a significant carrier-induced shift of the atomic motion about the site. In particular, in a perfect crystal, the time of stay is $h/6J$, while the characteristic distortion-producing time is $\omega_0^{-1}(h\omega_0/2E_b)^{1/2}$ for $h\omega_0 \gg \kappa T$ and $\omega_0^{-1}(\kappa T/2E_b)^{1/2}$ for $h\omega_0 \ll \kappa T$. The general theme of this result is that a small polaron will form if a carrier resides at a site for even a small fraction of a vibrational period $2\pi/\omega_0$, since typically $2E_b/h\omega_0 \geqslant 20$. If, however, the carrier can move between sites sufficiently rapidly so that the atoms do not have ample time to respond to its presence, the electron will be able to exist in a nonpolaronic state. Indeed it is the rapid intersite motion, rather than a weak electron–lattice coupling, which leads to the carriers in crystalline silicon and germanium being nonpolaronic.

It should be noted that these ideas also apply to defects. An electron in a shallow large-radius defect state induces minor displacements of the surrounding atoms, while a well-localized electron in a deep defect state usually induces significant atomic displacment, which lowers its energy. In the nonpolaronic large-radius situation the carrier may be thought of as spending so little time at a given site as to preclude its producing a substantial polaron effect. For a well-localized carrier in a deep defect state, the opposite situation prevails. Furthermore, an energy barrier and concomitant time delay can exist for forming the polaronic deformation associated with occupying the deep state. Thus a carrier may occupy a shallow state at a defect for a considerable time before a sufficient polaronic distortion pattern is established about the defect so as to localize severely the particle into a deep state. While the time delay need not always be very long ($\gg 10^{-12}$ s) and shallow states often exist without a deep-state counterpart, in some situations striking time delays have been observed [21]. Similar effects should be considered in relation to deep defect states in amorphous semiconductors.

In addition to the possibility of polaron formation, one can also consider the question of *bipolaron* formation. One then envisions the carrier-induced displacements being such as to make it energetically favorable for two carriers of like sign to reside together spatially. For the simplest small-polaron model, as illustrated in Fig. 2.6, the joint presence of the two particles produces a doubling of the polaronic energy lowering which each particle experiences. Thus the energy of two particles sharing a common polaronic well, relative to that of two well-separated small polarons, is $[-2(2E_b) + U] - [-2(E_b)] = -2E_b + U$, where E_b is the small-polaron binding energy and U the coulombic repulsion energy. If $2E_b > U$, bipolaron formation is favored over the formation of isolated small polarons. It has been suggested that such a phenomenon occurs in narrow-band organic [22] and inorganic semiconductors [23], in liquids [24, 25], and in amorphous semiconductors [26]. Nonetheless, bipolaron

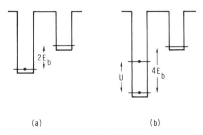

(a)   (b)

Fig. 2.6 The wells and energies associated with (a) simple small-polaron formation and (b) bipolaron formation. For a linear electron–lattice coupling, the net (polaronic) lowering of the system's energy, which includes the system's strain energy, is half that of the electronic levels.

formation, while an established concept, has as yet only been firmly demonstrated in a few systems [23].

There are essentially two models of the intrinsic equilibrium states of an amorphous semiconductor. The rigid (nonpolaronic) model envisions the disorder simply as introducing a fractionally small number of localized states at the band edges while leaving states in the bulk of the band extended. In this picture the mobility edge plays the role of the band edge in a crystalline semiconductor, with high mobilities being ascribed to charge transport in the extended states. On the other hand, the small-polaron model views charge carriers as relaxing into self-trapped states which lie within the absorption gap. Charge transport is then characterized in terms of hopping motion with rather low mobilities which increase with increasing temperature. While the properties of an amorphous semiconductor associated with each of these two views are the subject of the remainder of this chapter, some preliminary orienting comments are in order.

5. Applicability of the Models

One naturally expects amorphous semiconductors to be divisible into two classes contingent on which of these two models is appropriate. Indeed, the electrons in a-SiO_2 manifest the high mobilities expected of the rigid model [11]. However, the charge carriers in transition-metal-oxide glasses [27], the holes in a-SiO_2 [28] and in chalcogenide glasses [29], the electrons in a-As, and both species of carrier in a-Si and a-Ge manifest properties which are in reasonable accord with the small-polaron model [30, 31]. One may then ask, presuming that these assignments are correct, both generally and specifically, why small-polaron formation occurs in these materials.

From a very general point of view, one notes that disorder tends to impede and restrict motion, thereby facilitating self-trapping. Such behavior has been observed for exciton self-trapping in the mixed crystal $TlCl_x Br_{1-x}$ [32]. That is, while exciton self-trapping does not occur for either TlCl or TlBr, with sufficient mixing self-trapping is observed. A different type of example is that of sulfur; while electrons, but not holes, form small polarons in the crystalline solids, in S_8, upon melting with concomitant disordering, both species of carriers are found to self-trap [33]. Furthermore, in the ferromagnetic regime of the magnetic semiconductor EuO, oxygen vacancies serve as centers for shallow donor states which in the presence of magnetic disorder, introduced by passing to the paramagnetic regime, collapse into severely localized states [34]. These diverse examples illustrate the general principle: Disorder may produce sufficient localization to enable self-trapping and concomitant extreme localization to occur.

More specifically, one may speculate as to the origins of self-trapping in the previously enumerated examples of small-polaron formation in amorphous semiconductors. For holes in a-SiO$_2$ and d-band carriers in transition-metal-oxide glasses, the narrowness of the relevant bands is presumed to play the major role in providing the localization required for self-trapping. In the chalcogenide glasses, and to some extent in amorphous germanium, there is evidence for rapid localization of optically induced carriers, which implies that in these cases there is substantial localization *prior* to small-polaron formation [19, 35]. This evidence (to be discussed more fully subsequently) is of three types. First, unlike what transpires in high-mobility materials, excitation well above the gap does not efficiently yield (separated) charge carriers [36]; often optically generated charges remain in close proximity to one another and geminately recombine. The presence of such small electron–hole separation distances implies that there is extemely slow carrier motion, and rapid localization, before self-trapping. Also, the observation of low small-polaronic mobilities in photoconductivity experiments [37, 38] implies that free carriers relax very rapidly to the small-polaron state of much lower mobility. In other words, there is sufficient localization independent of self-trapping to reduce substantially the barrier and associated time delay for self-trapping, so that the transport which dominates the photoconductivity is of the low-mobility variety. Finally, optical excitation well above the absorption edge yields luminescence which remains un-Stokes-shifted long ($\gg 10^{-9}$ s) after excitation [39]. This contrasts with a typical high-mobility semiconducting crystal in which the generated carriers relax through a continuum of band states to the vicinity of the band edge in about 10^{-12} s. This result complements the observation of low carrier-generation efficiencies by implying that at least some of the high-energy excitation is generating excitons rather than charge carriers. Thus, consistent with significant localization, charge separation is very inefficient. All these results, as well, of course, as the evidence for small-polaron formation itself, suggest that there is a much greater tendency toward localization in (these) materials than is envisioned in the high-mobility picture.

Again, one may question why this is so. While the answer, or answers, are far from clear, preliminary observations can be made. The cases of small-polaron formation occur in situations in which the local orbitals associated therewith are directional (bonding, antibonding, p- or d-like), while the instance of high-mobility transport (electrons in a-SiO$_2$) has been associated with motion between somewhat spreadout s-like states. The energies of, and transfer between, directional local states will generally be more sensitive to the structural variations inherent in a glass than for s-like states. Thus systems which manifest small-polaron formation may simply possess much greater electronic disorder than nonpolaronic systems. Another question is whether some of the disordered structures are constructed as amalgamations of clusters of material which themselves provide structurally localized units that restrict electronic motion. Information about this morphological issue is currently sparse. Finally, there is extensive evidence in amorphous semiconductors (unlike crystalline materials) that a substantial fraction of the atoms of the material can readily move between nearly degenerate ($\lesssim 10^{-3}$ eV) distinct structural configurations [40, 41]. Often the distances involved in these atomic rearrangments ($\gtrsim 10^{-9}$ cm) are comparable to those characterizing small-polaron displacements. The presence of such latitude in the structure of an amorphous semiconductor can provide a predisposition toward small-polaron formation. If the local modes associated with these atomic displacements are sufficiently soft, the time a carrier can reside at a site before inducing substantial atomic displacments and small-polaron formation is especially short. In other words, it becomes less likely that a carrier can propagate through a noncrystalline solid without becoming self-trapped, simply because the effective strengh of the electron–lattice interaction is enhanced because of the presence of "soft modes." Regarding these notions, it should be realized that our knowledge of the noncrystalline structures and the theory of localization are not yet at a stage where they can be quantitatively applied to real materials.

2.4 ELECTRICAL TRANSPORT PROPERTIES

1. Conductivity

The most direct method of ascertaining the applicability of the rigid-atom (high-mobility) and small-polaron (low-mobility) models to specific systems is by determining the mobility directly from dc transport experiments. Presuming, as is often the case, that one species of carrier dominates the transport, the electrical conductivity σ is simply written as the product of the carrier density n, the charge of an individual carrier q, and the carrier's mobility μ: $\sigma = nq\mu$.

In the simplest version of the high-mobility scheme, the current is carried by a small number of high-mobility carriers which exist beyond the mobility edge. The equilibrium density of high-mobility carriers is given by

$$n = N(E_c)\kappa T \exp\left[-(E_c - \zeta)/\kappa T\right] \qquad (2.3)$$

where $N(E_c)$ is the density of states at the mobility edge E_c; the thermal energy is written as κT, and the Fermi energy as ζ. The mobility is presumed to be high ($\gg 1$ cm^2/V s) and to possess a much weaker temperature dependence than that of the carrier density. The conductivity in this model is thus often written

$$\sigma = \sigma_0 \exp\left[-(E_c - \zeta)/\kappa T\right] \qquad (2.4)$$

where σ_0 is nearly a constant. For a temperature-independent Fermi level, the conductivity has a simply activated temperature dependence.

In the simplest version of the small-polaron view, the equilibrium state is one in which both electron and hole carriers form small polarons. These (equilibrated) carriers reside within two very dense (vibrationally narrowed) bands of small-polaron levels. For a small-polaron band which is narrow compared with κT the polaron density of the predominant polaronic carriers, say electrons, is given by

$$n = N \exp\left[-(E_{sp}^c - \zeta')/\kappa T\right] \qquad (2.5)$$

where N is the density of potential carrier sites in the material ($\sim 10^{22}$ cm^{-3}), E_{sp}^c the characteristic energy of the band of the predominate carrier, and ζ' the Fermi level; the prime emphasizes that the position of the Fermi level generally differs from that of the rigid-atom model [42]. Typically, as illustrated in Fig. 2.7, the density of small polarons is much larger than that which would be associated with the rigid-atom scheme. This results from the larger prefactor and smaller activation energy which characterize the small-polaron situation.

The mobility of small polarons is generally rather low because self-trapped carriers can only move in response to an alteration in the positions

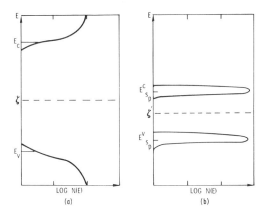

Fig. 2.7 The idealized densities of states for an amorphous semiconductor after lattice relaxation within (a) the rigid-atom and (b) small-polaron models.

of the atoms of the solid. In other words, the carrier-induced potential well which confines the carrier at a given site must be changed as a result of a shift of atomic positions before the carrier may move from the site at which it is confined. Since the small-polaron transfer integral is typically very much smaller than estimates of both the disorder energy and the energy uncertainty associated with scattering events, small-polaron motion in disordered solids is always described as phonon-assisted hopping motion [43]. That is, to calculate the mobility one asks for the rate, associated with atomic motion, that characterizes motion between specific carrier sites. In particular, for uncorrelated motion in a crystal the small-polaron mobility is written

$$\mu = (q/\kappa T)Ra^2 \tag{2.6}$$

where R is the elemental jump rate characterizing motion between adjacent sites separated by the distance a. In many instances in noncrystalline solids the disorder-related energetic fluctuation between adjacent sites appears to be sufficiently small so as to permit application of this formula [42, 43].

Calculation of the small-polaron jump rate is generally an involved matter. For details the reader is referred to the literature [44–50]. However, the physics of the jump process and the essential features of the results can be understood qualitatively from the following illustrative example. Consider the motion of a small polaron between two deformable diatomic molecules linked to other similar molecules of a molecular crystal. As illustrated in Fig. 2.8a, because of the degeneracy associated with the occupation of either of the two molecules, the carrier can move between the pair of molecules when the atoms make a suitable readjustment. That is, the atoms of each diatomic molecule must "tunnel" between displaced and undisplaced configurations. Since the matrix element for this intrasite atomic tunneling is small, the intermolecular jump rate is also small. However, as depicted in Fig. 2.8b, motion can be facilitated when a vibration-related fluctuation of atomic configurations occurs, which reduces the disparity between the initial and final atomic configurations of the tunneling event. Such fluctuations become more prevalent as the temperature is raised. Thus the small-polaron jump rate increases with increasing temperature. Furthermore, as shown in Fig. 2.8c, for a suitable fluctuation, termed a coincidence event, which requires still more energy, there is no disparity between initial and final molecular configurations. The transfer is then no longer impeded by the requirement that the atoms of the molecules tunnel between different configurations. For electronic small-polaron hopping the motion in this regime is simply activated with the activation energy W_H, the minimum energy which must be supplied to

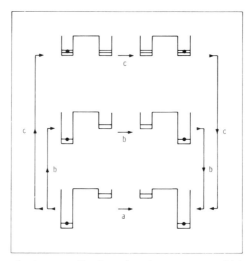

Fig. 2.8 The predominant small-polaron hopping processes at (a) low, (b) intermediate, and (c) high temperatures.

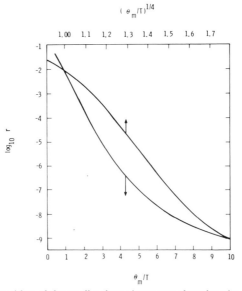

Fig. 2.9 The logarithm of the small-polaron jump rate plotted against reciprocal temperature in units of the temperature corresponding to the maximum-energy phonon with which the carrier interacts appreciably, θ_m. The curve is also plotted against the fourth root of θ_m/T. (See Emin [45] for details.)

the equilibrated pair of molecules to create a coincidence event. Thus, as depicted in Fig. 2.9, the small-polaron jump rate rises with temperature in a nonactivated manner until this classical regime, characterized by the occurrence of coincidence events, is encountered. Beyond this temperature the curve assumes an Arrhenius character. The transition from activated to nonactivated behavior typically occurs at some fraction $(\gtrsim \frac{1}{3})$ of the characteristic photon temperature of the material [45–47].

In the high-temperature regime the jump rate may be written as the product of the rate at which coincidence events occur and the probability P that the charge carrier will avail itself of the coincidence event and move between sites [43]. Assuming a random occurrence of coincidence events one writes the rate as

$$R = \left[\nu_{\mathrm{vib}} \exp(- W_{\mathrm{H}}/\kappa T) \right] P \tag{2.7}$$

where ν_{vib} is the characteristic vibrational frequency of the system. If the electronic transfer integral is sufficiently small (typically $\lesssim 10^{-2}$ eV), P is both much less than unity and proportional to $|J|^2$; the hopping is then termed nonadiabatic. While this regime is often discussed, it is the complementary (adiabatic) regime [51], characterized by larger transfer integrals and a probability factor of order unity, which is often the regime of physical interest. For adiabatic hopping in the high-temperature regime one obtains a simple formula for the dc conductivity:

$$\sigma = \left(Nq^2 a^2 \nu_{\mathrm{vib}}/\kappa T \right) \exp\left[-(E_{\mathrm{sp}} - \zeta' + W_{\mathrm{H}})/\kappa T \right] \tag{2.8}$$

where the prefactor has a room temperature value of about 10^3 $(\Omega\,\mathrm{cm})^{-1}$.

2. Thermoelectric Power

To ascertain whether the conductivity of a given system is attributable to high- or low-mobility carriers requires methods of separately obtaining the carrier density and mobility. Is the current carried by a high density of low-mobility carriers or by a low density of high-mobility carriers? Several schemes can be employed. These involve measurements of the thermoelectric power, the Hall coefficient, the photoconductivity, and the field effect.

A thermoelectric power experiment involves measuring the emf generated across a material in response to a small temperature gradient. Specifically, the thermoelectric power S is the ratio of this emf to the temperature gradient. It is related to the Peltier heat Π by the Kelvin relation $S = \Pi/qT$. The Peltier heat is the energy transported with a charge carrier as it moves through a material. In the simplest models of both the high- and low-mobility situations [9],

$$S = (\kappa T/q)(\delta \ln n/\delta T)_{\zeta} \tag{2.9}$$

where the sign of the thermoelectric power is given by the sign of the carrier q since $(\partial \ln n / \partial T)_\zeta$ is positive. Specifically, in the high-mobility situation the thermoelectric power is typically given by

$$S = \frac{1}{qT}\,(E_c - \zeta + A\kappa T) = \frac{\kappa}{q}\left(\frac{E_c - \zeta}{\kappa T} + A\right) \qquad (2.10)$$

where A is a factor of order unity. Here it is presumed that the mobility edge is independent of temperature. If it is not, additional terms generally enter into the square brackets [52].

In the simplest version of small-polaron transport none of the hopping energy is transported with the carrier as it hops, since the hopping energy is absorbed (and emitted) equivalently at the initial and final sites of a hop [53]. In this case,

$$S = \frac{1}{qT}\,(E_{sp} - \zeta' + A'\kappa T) = \frac{\kappa}{q}\left(\frac{E_{sp} - \zeta'}{\kappa T} + A'\right) \qquad (2.11)$$

where A' vanishes for transport among states with a characteristic width W_{sp} which is much smaller than the thermal energy κT [54]. However, A' rises with decreasing temperature to a peak at $\kappa T \approx W_{sp}$ and then falls to a constant value ~ 1 when $W_{sp} \gg \kappa T$; typically the maximum value of A' is less than or comparable to 10. By comparing the temperature dependence of the conductivity with that of the thermoelectric power, one can determine if there is a difference between the conductivity activation energy and the energy characterizing the slope of a plot of S versus T^{-1}. If there is a significant difference, small-polaron hopping is implied, with the energy difference being attributed to the hopping activation energy W_H. Furthermore, with the measured energy separation $E_{sp} - \zeta'$ one can use Eq. (2.5) to estimate the carrier density. With this and the conductivity known, the mobility can be determined. Finally, using both the mobility and the hopping energy, the mobility prefactor can be obtained.

Conductivity and thermoelectric power data on chalcogenide glasses [42, 55], a-Si [56, 57], a-Ge [31, 38], and a-As [58], as well as on transition-metal-oxide glasses [59], all show significant energy differences which can be interpreted as being small-polaron hopping activation energies. For example, in the chalcogenide glasses the values of W_H range from about 0.15 eV for As_2Te_3 to about 0.6 eV for As_2S_3 [55].

Alternative explanation of these results in terms of the high-mobility picture have also been sought. One scheme is to assume that the conduction involves transport in two bands of states, such as extended states above the mobility edge and localized states below the edge, in which both modes have nearly equal partial conductivities [60–62]. Then the parameters of the two-band formulas,

$$\sigma = \sigma_1 + \sigma_2 \tag{2.12}$$

and

$$S = (S_1\sigma_1 + S_2\sigma_2)/(\sigma_1 + \sigma_2) \tag{2.13}$$

can be chosen to match the data (with $\sigma_1 \approx \sigma_2$) over a limited temperature range. An explanation such as this, which depends on several parameters lying within rather stringent limits, seems inappropriate for such a commonly observed phenomenon. Another attempt, which is only applicable to situations in which there is considerable bending of the $\ln \sigma$ versus T^{-1} and S versus T^{-1} curves, views the energy differences as arising from the temperature dependences of σ_0 and $A\kappa T$, in Eqs. (2.4) and (2.9) [63], respectively. While the conductivity and thermoelectric power curves of doped amorphous Si [57] by themselves may admit such an explanation, in other materials there is too little curvature of the $\ln \sigma$ versus T^{-1} and S versus T^{-1} curves to be consistent with such a view [55, 56, 59]. In all cases, one must look for corroborating evidence for each explanation from other studies.

3. Hall Effect

Hall effect studies are a standard tool for studying the mobility of charges in solids. In these experiments one applies a magnetic field perpendicular to the current flow and observes the resulting deflection of the current. The measured quantity is the Hall coefficient R_H. The product of the Hall coefficient and the conductivity is the Hall mobility: $\mu_H = \sigma R_H$. Physically, the Hall mobility is proportional to the transverse current induced by the Lorentz force. For free electrons $\mu_H = \mu$. However, for electrons in a solid, the Hall mobility and the conductivity mobility generally differ from one another. In the case of high-mobility motion in crystals, the two mobilities just differ by a factor, termed the Hall factor, which is typically on the order of unity. Thus in these instances, the Hall mobility is taken to be a measure of the mobility which enters into the conductivity expression. Furthermore, in high-mobility crystals with single-band conduction, the sign of the Hall coefficient and the sign of the thermoeletric power provide two independent determinations of the sign of the carrier. That is, the sense in which the carrier is deflected by a magnetic field is used to determine the sign of its charge.

As with the conductivity mobility, there are qualitative differences between the Hall mobility associated with small-polaron hopping and that associated with the motion of extended-state carriers. A basic reason for the differences is that the electronic wave function of a small polaron is severely localized while that of an extended-state carrier has a coherence length which is much larger than an interatomic spacing. Consequently, the motion of an extended-state carrier is insensitive to atomistic detail. Indeed, it is well known that the similarity of the motion of excess charges in a solid to that of free charges results from the coherence length of the charges' wave functions being much greater than an intersite separation. The small-polaron Hall mobility is, however, quite distinct in that it is sensitive to the geometric arrangement of sites involved in the jump process and the symmetry of the electronic orbitals between which motion occurs.

Central to an understanding of the Hall mobility associated with hopping motion is the notion that the application of a small magnetic field can only be effective in altering the motion of a carrier when the carrier is confronted with a choice between hopping to each of two or more sites [64]. In other words, for high-temperature hopping motion, the Hall effect is associated with the occurrence of coincidence events which involve at least two potential final sites in addition to the initially occupied site [65]. As a result, the small-polaron Hall mobility is characterized by an activation energy which is different (usually considerably smaller) than that of the conductivity mobility W_H [30]. Thus the small-polaron Hall mobility is generally low (< 1 cm^2/V s) and is activated. However, at times the activation energy is not sufficiently large for the activated temperature dependence to dominate the algebraic decrease with increasing temperature of the preexponential factor in the Hall mobility expressions. Such a situation typically produces a very weak temperature dependence for the Hall mobility in which the Hall mobility may even fall with increasing temperature. To illustrate this point, three examples of the conductivity and Hall mobilities are plotted against reciprocal temperature in Fig. 2.10.

A striking feature of small-polaron transport is the possibility of Hall effect sign anomalies [30]. In these instances the magnetic field deflects excess electrons in the sense of free positively charged particles and deflects holes in the sense of free negatively charged particles. The occurrence of these anomalies manifests the severely localized character of the carriers. As described in detail elsewhere, the local character of the transport manifests itself in the dependence of the sign of the Hall effect on such local properties as the geometric arrangement of the sites between which the carrier hops and the symmetry of the orbitals involved in the hopping [30, 66]. In particular, sign anomalies can arise when linking of the centroids of the orbitals between which the carrier can directly hop yields a construction composed of predominantly odd-membered rings. Such constructs are consistent with various structural models of amorphous silicon,

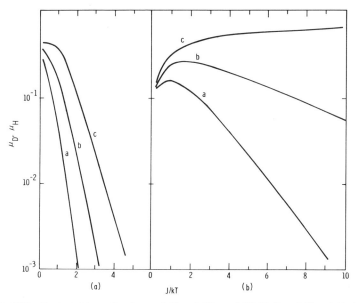

Fig. 2.10 Typical (a) small-polaron drift mobility and (b) Hall mobility plotted against reciprocal temperature, in units of the electronic transfer integral, for three values of the physical parameters. (See Emin and Holstein [51] for details.)

germanium, and arsenic. Electron hopping between antibonding orbitals then yield a p-type Hall effect, while holes moving between bonding orbitals produce an n-type Hall effect. In heteropolar materials, such as the chalcogenide glasses and a-GaAs, the relevant construction not only depends upon the geometric structure and the nature of the local orbitals but requires designating the sites between which each species of carrier moves. For example, the holes in a-As_2Te_3 are thought to move among three-membered rings of chalcogen atoms, which comprise the base of the pyramidal As–Te_3 units of which the glass is composed. For a trifurcated coorientation of the lone-pair p-orbitals, hole small-polaron hopping then yields an n-type Hall effect. As another example consider a-GaAs. Although a-GaAs may be composed of six-membered rings in which Ga and As atoms alternate positions, the electrons and holes may be viewed as each moving between like atoms, Ga and As, respectively. The construction then yields three-site triangular arrangements of like atoms.

The low, anomalously signed, and often thermally activated Hall mobilities observed for electron and hole transport in numerous amorphous semiconductors, a-Si, a-Ge, a-As, and many chalcogenide glasses, are in general accord with the predictions of the small-polaron picture [30]. However, interpretations of these experiments in terms of the competition among various charge transport modes have also been advanced [60–62]. These schemes often do not account for the sign anomalies [67] and almost always require various adjustable parameters to lie within rather restrictive limits [68–70].

In several ternary glasses, TlTeAs [71] and CdGeAs [72], of unknown structure, the Hall mobilities are reported to be low with the Hall effect signs being normal. Unlike the simpler systems, in these systems significant changes in the transport (such as even the Hall effect sign) with composition and temperature suggest that multiband conduction may be occurring.

4. Photoconductivity

Steady-state and transient photoconductivity measurements are generally regarded as providing a method for ascertaining the magnitude and temperature dependence of the mobility of a charge carrier. While this is a well-accepted procedure, there are some interesting complications which can enter into the interpretation if small-polaron formation occurs. For example, consider the standard view of the low-intensity regime characterized by the density of optically generated carriers being much less the equilibrium density. Here for an intrinsic semiconductor the photoinduced conductivity for one species of carrier of mobility μ_D may be written

$$\Delta\sigma = \delta n e \mu_D = G\tau e \mu_D \qquad (2.14)$$

where Δn, the photoinduced excess carrier density, is the product of the carrier generation rate G and the photoconductivity decay time τ. By measuring the transient behavior of the photoconductivity the decay time may be determined. With a knowledge of the energy distribution of incident photons, as well as the absorption coefficient and quantum efficiency of the material as functions of the photon energy, the carrier generation rate may be determined. A knowledge of $\Delta\sigma$, G, and τ readily yields the mobility μ_D.

Low-mobility semiconductors may not only be distinguished from high-mobility semiconductors by the carriers forming small polarons but also by their low quantum efficiency η. While the quantum efficiency for super-band-gap excitation in high-mobility semiconductors is typically comparable to unity, in low-mobility materials it may be many orders of magnitude smaller. As will be discussed again in subsequent sections, this low quantum efficiency is associated with an optically generated electron–hole pair recombining with each other rather than separating and contributing to the photoconductivity. Such recombination is termed geminate recombination. The complementary process, associated with a pair avoiding geminate recombination and contributing to the photoconductivity, is characterized by a rate which increases strongly with temperature. That is, here the quantum efficiency increases strongly with temperature. If one were unaware of this phenomenon (and treated η as a nearly temperature-independent quantity of order unity), the mobility determined from the photoconductivity would both be artificially low and contain an artificial factor which rises with increasing temperature.

In materials in which the predominant charge carriers form small polarons, the mobility entering into the photoconductivity may differ from the simple small-polaron mobility. One reason for this is the time delay for small-polaron formation. That is, the carrier will exist in a nonpolaronic state for some fraction f of its lifetime before small-polaron formation occurs. In a simple two-state picture, the photoconductivity mobility may be written

$$\mu_D = \mu_{np} f + \mu_{sp}(1 - f) \qquad (2.15)$$

where μ_{np} and μ_{sp} are the mobilities associated with the nonpolaronic and small-polaronic states, respectively. Thus the mobility determined from photoconductivity experiments for systems in which equilibrated carriers form small polarons may reflect the mobility of the precursor (non-polaronic) state, rather than that of the small polaron, if the lifetime of this state t_{np} is a sufficiently large portion of the net lifetime. If this is the case, the temperature and photon-energy dependences of the apparent mobility μ_D are affected by the fact that the time t_{np} that the carrier resides in the nonpolaronic state decreases with temperature as the rate of its conversion to the small-polaron state increases. In addition, this conversion rate tends to be a decreasing function of the energy of the exciting photon, presuming that increasingly delocalized prepolaronic states are populated. However, the strength of this dependence may vary significantly from system to system. Alternatively, the observation of a small-polaronic mobility in photoconductivity experiments places some limits on the lifetime of the nonpolaronic component $t_{np} \ll [\mu_{sp}/(\mu_{np} + \mu_{sp})]\tau$.

Another effect which may produce significant discrepancies between the mobilities determined by photoconductivity experiments and by dark-transport measurements is related to the relaxation of the excited carrier after it forms a small polaron: The energy distribution of occupied small-polaron states after photoexcitation will generally differ from the thermo-dynamic distribution. Thus, if the small polaron initially occupies sites of higher energy than would be consistent with an equilibrium distribution, the photoconductivity will tend to be associated with a greater number of hops downward in energy than is the dark conductivity. As a result the temperature dependence of the mobility will be weakened. This effect is most noticeable at low temperatures, typically $\lesssim \frac{1}{3} T_{Debye}$, where the pre-dominance of such downward hopping, associated with the spontaneous emission of phonons, yields a nearly temperature-independent mobility [45–47]. Such a situation has been characterized as a degrading polaron [38].

Photoconductivity experiments on amorphous germanium and amorphous arsenic telluride have been interpreted in terms of a small-polaron hopping motion [37, 38]. In these works there is evidence for the relaxation effects mentioned above.

5. Hopping between Defects

Various amorphous semiconductors appear to possess sufficiently large densities of defects so that conduction associated with direct motion between defect states dominates the transport over significant temperature ranges. For example, states associated with dangling bonds in unhydrogenated amorphous silicon can exist in sufficient densities, $\sim 10^{19}$ cm^{-3}, so that hopping between these states dominates the transport below room temperature [73]. While this is somewhat similar to shallow-impurity conduction in crystalline semiconductors, such as that due to motion between phosphorus donors in crystalline silicon, there are important fundamental differences between the two situations. The standard impurity conduction problem addresses the low-temperature ($\ll 10$ K) hopping of carriers between large-radius (~ 50–100-Å) impurity states which, because of their large spatial extent, interact very weakly with the lattice vibrations [46]. As such, each of the phonon-assisted transitions between impurity states (hops) primarily involves only the emission or absorption of a single low-energy phonon of energy well below the Debye energy [74]. In amorphous semiconductors the hops occur between deep-lying, well-localized (radii ~ 5 Å) defects at very much higher temperatures: above ~ 100 K in a-Ge, ~ 300 K in a-Si, and ~ 500 K in vanadate glasses. As a result of their severe localization, the carriers interact strongly with the atomic vibrations [45, 46]. Each of the interdefect hops therefore generally involves the absorption and/or emission of very many phonons. Thus, while the energies exchanged between the carrier and the atomic vibrations in a typical hop are 10^{-3}–10^{-4} eV for impurity conduction [74], for defect hopping in amorphous semiconductors they are about three orders of magnitude larger [73, 75].

The rates which characterize a nonadiabatic hop between the initial and final sites are given by

$$R_{i,f} \propto |J_{i,f}|^2 \exp\left[-(E_f - E_i)/2kT \right] F\left[T, (E_f - E_i) \right] \qquad (2.16)$$

in the strong coupling multiphonon regime, and by

$$R_{i,f} \propto |J_{i,f}|^2 \exp\left[-(E_f - E_i)/2kT \right] \operatorname{csch}\left[|E_f - E_i|/2kT \right] \qquad (2.17)$$

in the low-temperature weak-coupling single-phonon regime of relevance to the impurity conduction problem [45–47]. Here $|J_{i,f}|^2$ is the absolute square of the transfer integral associated with a transition between the initial and final sites, characterized by the energies E_i and E_f, respectively. With the exception of rather low temperatures and very large energy disparities, $F[T,(E_f - E_i)]$ may be approximated by its form at zero-energy disparity, $E_f = E_i$. The temperature of this function is then just that of the jump rate between degenerate sites illustrated in Fig. 2.9. The dependence of the jump rate on the energy disparity is then simply contained in the factor $\exp[-(E_f - E_i)/2kT]$. In the weak-coupling regime with $E_f - E_i \gg 2kT$, the temperature dependence of the jump rate is given simply by the factor $\exp\{-[(E_f - E_i) + |E_f - E_i|]/2kT\}$. Here the rate varies as $\exp[-(E_f - E_i)/kT]$ for hops upward in energy while becoming temperature-independent for hops downward in energy.

To compute the conductivity associated with hopping in either case involves finding the most efficient paths along which a charge can move through a disordered array of defect or impurity sites. The ease of any given hop is determined by two factors: the transfer integral and the energy disparity. The factor $|J_{i,f}|^2$ of the jump rate expressions tends to fall as one considers pairs of sites which are increasingly separated spatially. This effect favors hops between sites which are in close proximity to one another. However, the jump rates also place a premium on avoiding hops upward in energy to sites of especially high energy. Since the probability of finding a low-energy final site increases as one considers increasingly large numbers of sites, lying within ever-larger distances from the initial site, there is competition between the two factors. The transfer integral factor encourages short hops, while the energy disparity dependence tends to favor long hops. As the temperature is lowered, the energy disparity factor grows in relative importance, since $(E_f - E_i)/2kT$ grows. The conductivity is then determined by an increasingly large proportion of long hops with ever-smaller energy disparities. Concomitantly, the temperature depence of the conductivity becomes milder. Thus this phenomenon yields a conductivity which rises with temperature in an non-Arrhenius manner, up to a maximum temperature at which the range of the hops which determine the conductivity approaches the average intersite separation [76].

As a particular model it has been assumed that (1) the transfer integral falls off exponentially with intersite separation and (2) the density of states of the involved sites is independent of energy. Then it has been argued that this variable-range hopping yields a conductivity which varies as $\exp[-(T_0/T)^{1/4}]$ [76], where T_0 is a constant which depends on the density of sites. While for many amorphous semiconductors the logarithm of the conductivity is linear when plotted against $T^{-1/4}$ over some temperature range, the spread of temperatures investigated is often not great enough to permit one to distinguish between various nonactivated temperature dependences [77, 78]. For example, even measurements between 25 and 300 K represent less than a factor of 2 change in the abscissa.

It has been pointed out that the factor $F(T)$ manifests a nonactivated temperature dependence similar to that predicted by the variable-range hopping approach [45]. Here the freezing out of multiphonon processes rather than the freezing out of short, high-energy hops produces the non-Arrhenius temperature dependence of the conductivity. In this circumstance the maximum temperature at which the nonactivated temperature dependence is observed should crudely scale with the characteristic phonon temperature of the material. The available evidence is consistent with this notion [73, 75]. A more direct test is to look for a change in the temperature dependence of the conductivity as the density of sites is altered by hydrogenation or annealing of the material. At least in the simplest models, the temperature dependence of variable-range hopping should be altered by such treatments, while the multiphonon effect should not. Here the evidence indicates the predominance of the multiphonon effect [31, 36, 56]. Another procedure which is sometimes applicable is to compare the magnitude of the characteristic energy disparity determined from thermoelectric power measurements with that required for a variable-range hopping interpretation of the conductivity data. In amorphous silicon and the vanadate glasses the energy disparities required by the variable-range hopping theory are much larger than those determined from thermoelectric power experiments [31, 56, 59, 75]. Another scheme, based on the assumption that thin films of the amorphous materials are homogeneous, is to investigate the conductivity of the semiconductor films as their thickness is reduced to distances comparable to intersite separations. In the variable-range hopping picture, as one constricts the conductivity paths, higher-energy hops are forced and the conductivity is both reduced in magnitude and increased in temperature dependence. For a homogeneous material such changes in sample size should not affect the multiphonon effect. Experiments on amorphous silicon and germanium indicate a change in temperature dependence in accord with the variable-range hopping picture [79]. It should be noted that a similar effect can be associated with the increasing role of surface states as the film thickness is reduced. The Fermi level of the material will generally shift away from the bulk states as the film thickness, and hence the ratio of bulk to surface states, is reduced. As a result, the difference between the Fermi level and the energy of the bulk defect states is increased. This produces an increase in the temperature dependence of the conductivity as the film thickness is reduced. The disparities between the various experiments and their inter-pretations have yet to be unambiguously resolved.

2.5 OPTICAL PROPERTIES

2. Absorption and Carrier Generation

The absorption spectra of amorphous semiconductors generally mani-fest gaps much like those characteristic of crystalline semiconductors. However, the absorption of a super-band-gap photon need not produce effects similar to those which characterize such well-known semiconduc-tors as crystalline silicon. Rather, as in transport, two quite distinct situations can be envisioned. The first is an itinerant picture like that characteristic of high-mobility crystalline semiconductors. An alternative is a local-type picture much like that of a low-mobility molecular crystal in which the electronic excitations are sufficiently immobilized so as to become self-trapped. These two situations will now be briefly described.

In either case the Franck–Condon principle stipulates that the atoms of the solid will remain essentially fixed during the absorption of the super-band-gap photon. The distinction between these two situations lies in the nature of the optically induced excitation and the subsequent response of the atoms of the material. Thus, for example, since polaronic relaxations occur only after excitation, polaronic features such as the small-polaron band narrowing do not manifest themselves in the absorption spectra.

In the itinerant picture, the absorption of a super-band-gap photon typically produces an electron–hole pair with sufficient kinetic energy to separate. For example, a super-band-gap photon which imparts 0.1 eV to the relative motion of an electron–hole pair will, presuming the carriers behave as free particles, convey sufficient momentum to the carriers to allow them to separate about 2000 Å in only 1 ps. This separation distance is much larger than the coulombic capture radius $r_c = e^2/\epsilon kT$ (where ϵ is the dielectric constant) at all but very low temperatures; typically $r_c \approx$ 50 Å and $\epsilon = 12$ at room temperature. Thus, the absorption of a super-band-gap photon produces an electron and hole which break apart and contribute to the photoconductivity. In addition, the rate of energy loss of excited itinerant carriers is sufficiently fast, $\approx 10^{12}$ eV/s, so that excited itinerant carriers relax to their respective band edges in times less than or comparable to 1 ps. Thus, high-energy states in a band only remain populated for a very brief time period after excitation.

In a localized, molecular-solid-type picture, the absorption of a super-band-gap photon may not produce electrons and holes which readily separate from one another [80]. One may envision the incident photon as only promoting a charge of a localized entity to an excited state. If the transfer integrals and the differences in site energies are such as to confine the carrier to a restricted region, the carriers will then not readily separate. As a result, the photons will produce excitonlike entities. However, unlike the situation for Wannier-type excitons in crystalline silicon, but similar to the situation in molecular crystals, the band of exciton states may be quite wide, on the order of several electron volts. Thus, much of the observed absorption may be associated with creating such excitations. While one can only speculate on the microscopic details of such a situation, its experimental manifestations are quite clear. First, rather than the absorption of super-band-gap photons efficiently producing charge carriers, the carrier generation efficiency will be extremely low as most carriers geminately recombine. Second, since the relaxation of the excited molecular-type entities can be relatively slow ($\sim 10^{-6}$ s is typical of molecules) [80], super-band-gap excitation may produce observable super-band-gap luminescence. Third, because the self-trapping of severely constrained particles occurs rapidly [9], the mobilities of the photoinduced carriers produced will manifest low small-polaronic mobilities rather than the higher mobilities which may be associated with the carriers before self-trapping occurs.

Many amorphous semiconductors manifest at least some of these phenomena. These materials include a-Ge [38, 81] and chalcogenide glasses [36, 39, 82]. It should be noted that the failure to obtain charge separation could also be associated with the presence of a high density of electron and hole traps which trap *both* species before they can separate from one another. If either charge species remains mobile, charge separation could still occur.

2. Midgap Absorption

Some absorption can also occur within the forbidden gap. This can be associated with the excitation of carriers out of traps. An analogous phenomenon is the absorption associated with the excitation of a self-trapped charge out of its induced potential well [30, 43]. The distinctive feature of this small-polaron absorption is that the simplest theory predicts the peak of the absorption band to occur at an energy nearly equal to four times the high-temperature small-polaron hopping activation energy [83]. Absorption bands about these values have been observed in chalcogenide glasses [82] and in a-Ge [81]. The magnitude of the absorption bands in either case is proportional to the density of self-trapped or trapped charges.

3. Optically Induced Properties

With sufficiently intense optical excitation at low temperatures, the density of self-trapped or trapped charges can be increased to a level at which their properties can be readily studied. In particular, in various amorphous semiconductors metastable optically induced ESR signals and midgap absorption bands have been induced [82]. It has been suggested that these induced properties are associated with trapped [2, 3] as well as self-trapped carriers [84]. For example, the two induced ESR signals in a-As_2Se_3 have been associated with electrons localized on As atoms and holes localized about Se atoms [82]. The question is whether these sites are the centers of defect trapping or of self-trapping. One potential means of obtaining a resolution of this question is by attempting to saturate the traps. Presumably, when the traps become filled, additional illumination will no longer augment the induced ESR and induced absorption. While the absorption due to small polarons has a tendency to saturate at high densities, because of the increased probability of recombination, small-polaron saturation should occur at higher densities than the trap densities suggested for materials such as chalcogenide glasses, $\sim 10^{17}$ cm^{-3} [2, 3]. Attempts to investigate the saturation suggest that rather high densities, $\sim 10^{20}$ cm^{-3}, can be induced [85]. However, it must be cautioned that some of the assumptions made in extracting these estimates from the experimental data may be inappropriate in some cases. Thus, this issue requires further investigation.

It is interesting to note that the optically induced midgap absorption and ESR signals can be bleached by excitation within the midgap absorption band. This bleaching is rather inefficient; in chalcogenide glasses $\sim 10^2$–10^3 photons must be absorbed for each absorption center (and concomitant pair of ESR centers) to be destroyed [82]. In the small-polaron view the bleaching just corresponds to optically inducing the self-trapped carrier to move [84]; in the simplest small-polaron approach the absorption of each midgap photon induces one small polaron to hop to an adjacent site. Thus, many random hops must occur before oppositely charged small polarons hop close enough together to recombine and thereby have their ESR and absorption properties bleached. In the trapping picture, the low bleaching efficiency may be associated with a substantial probability of an optically freed carrier falling back into its original trap; such a picture would characterize traps which are charged when vacant and neutral when filled. An interesting finding is that the bleaching efficiency is uniform throughout the entire width (> 0.5 eV) of the midgap absorption band [82]. In the small-polaron picture this is in accord with the prediction that a photon absorbed anywhere within the absorption band produces a hop between adjacent sites; there is no dependence of the bleaching efficiency per absorbed photon on the photon's energy [84]. However, with the view that high-mobility carriers are being freed from charged traps, one expects the absorption of increasingly high-energy photons to enchance the rate at which the defects are ionized and bleaching occurs.

A variety of amorphous semiconductors display cw luminescences which are considerably Stokes shifted [86]. In the chalcogenide glasses this luminescence is shifted by about 1 eV. This implies that the radiative recombination occurs with the aid of deep recombination centers or via the recombination of small polarons. Furthermore, in the chalcogenide glasses the temperature dependence of the luminescence indicates that there is a competing nonradiative recombination process which dominates the recombination at essentially all temperatures. This is consistent with the presence of severely localized centers which interact strongly with the atomic vibrations [45]. In materials such as hydrogenated amorphous silicon there appear to be various competing recombination processes whose relative importance is determined, at least in part, by the preparation of the material. Since the situation in the chalcogenide glasses appears, at least at present, to be least complex, the remainder of the discussion of luminescence will focus attention on these glasses.

4. Transient Luminescence and Midgap Absorption

To better determine the relevance of various models to the luminescence, one can study the evolution of the luminescence after initial excitation. Recently such experiments have been performed on a-As_2S_3 [39, 87, 88]. The data appear to be readily interpretable in terms of the simplest of small-polaron models [19, 88]. Hence, in what follows, the model will be described and the principal experimental findings indicated.

In this localized picture, the unrelaxed electronic excitation generated by a super-band-gap photon is taken to be characterized by a distribution of charge separation lengths. If the resulting electron and hole remain sufficiently close together so that their wave functions overlap substantially, they are regarded as a neutral entity, an exciton. Otherwise, whether or not they geminately recombine, charges whose wave functions do not overlap substantially are viewed as separated, or metastably separated [84], charges. Since the interaction of an exciton with the atoms in its immediate vicinity is typically much weaker than that of a single carrier, the exciton is taken to be characterized by a much longer time delay toward small-polaronic relaxation than are separated charges. In this discussion, consistent with studies on chalcogenide glasses [89], the self-trapping of separated charges is presumed to occur in about 1 ps.

The kinetics of the hopping of separated electronlike and holelike small polarons from their positions at creation to those from which they recombine reflects itself in the time dependence of the luminescence and induced absorption. To understand the dynamics of the recombination of the

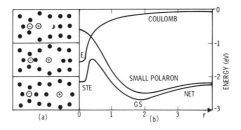

Fig. 2.11 (a) The equilibrium displacement patterns for three separations of electron and hole small polarons. (b) The corresponding coulombic, small-polaron, and net energy of the polaron pair plotted as a function of their separation.

electron and hole small polarons involves considering the energy of the system as a function of the pair separation [19]. As illustrated in Fig. 2.11a, in addition to their coulombic attraction, they interact via the overlap of their atomic displacement patterns. The coulombic, polaronic, and net potential energies of the pair as functions of their separation r, in units of the interatomic spacing, are plotted in Fig. 2.11b; for like-signed small polarons one simply inverts the energy scale. The three pair-potential minima corresponding to the formation of an exciton (E), a self-trapped exciton (STE), and the ground state of separated electron and hole small polarons (GS) are indicated on the figure.

Carriers which do not separate ($r_{initial} < 1$) after excitation are viewed as forming excitons which ultimately self-trap to STE. Charges which separate ($r_{initial} > 1$) rapidly ($\sim 10^{-12}$ s) relax to form small polarons at $r > 1$. They then tend to hop toward their ground state before recombining. Those that radiatively recombine before reaching GS will luminesce at a higher energy than those which reach GS. Furthermore, since the overlap and the matrix elements for recombination decrease with increasing separation, the luminescence will tend to shift to lower energy with time as the small-r pairs recombine first.

As summarized in Table 2.1, after $\sim 10^{-12}$ s, three luminescence bands result. At the highest energy lies the simple-exciton luminescence. Since there is little lattice relaxation about the localized exciton, it manifests a minimal Stokes shift. In chalcogenide glasses, the parameters are such that the recombination is primarily radiative. The extreme smallness of the competing nonradiative rate results from the fact that the recombination energy (~ 2–3 eV) greatly exceeds the exciton–lattice coupling energy $[\Delta E_E \ll \Delta E_{STE} \sim (E_E - E_{STE}) < 1$ eV] [39, 46, 87]. A lower-energy luminescence, manifesting a Stokes shift, is due to the self-trapped exciton. As indicated above, the nonradiative recombination associated with the STE, while larger than that for the simple exciton, will still not be able to compete effectively with the radiative recombination. For both excitons

TABLE 2.1

Low Temperature Properties of the Three Intrinsic Luminescence Bands

Source	Strokes shift	Recombination	Decay rate
Exciton	Small	$R_{rad} \gg R_{NR}$	$\sim T$ independent
Self-trapped exciton	Moderate	$R_{rad} \gg R_{NR}$	$\sim T, E$ independent
Small polaron	Large; shifts with time	$R_{rad} \ll R_{NR}$	Decreases with T and E; long tail in time

the luminescence decay rates at low temperatures will manifest little temperature dependence. However, at higher temperatures (comparable to the phonon temperature) thermally assisted tunneling through the self-trapping barrier from E to STE will enhance the STE luminescence at the expense of that of the simple exciton. Finally, the lowest-energy luminescence is associated with the recombination of separated small polarons. As indicated above, this Stokes shift increases with time as more luminescing carriers converge on their minimum energy separation. Furthermore, calculations show that, as a result of reduced electronic energy separation and enhanced electron–lattice coupling strength, nonradiative recombination dominates recombination at the ground state and in its vicinity. In addition, reflecting both the predominance of nonradiative recombination and the thermally assisted motion of the separated charges, the small-polaron population and its associated luminescence will shift and decay with increasing rapidity as the temperature is raised above a fraction ($\sim \frac{1}{3}$) of the phonon temperature.

An important feature of small-polaron hopping is that the jump rates depend strongly on the energy differences between the initial and final sites [46, 84]. For example, with the hole–lattice coupling determined for a-As_2Se_3, a change in energy difference from 0.2 to 1.2 eV produces an increase in the downward (spontaneous emission) jump rate of ten orders of magnitude [84]. As a result, pairs with separations on the relatively steep portions of the small-polaron $E-r$ curve hop toward the ground state rather quickly (10^{-8} s per hop), while those that are sufficiently well separated so as to experience small energy gradients require extremely long times to move ($\sim 10^3$ s per hop for $r \geqslant 4$). These metastably separated small-polaron pairs contribute a long, slowly decaying (nonexponential) tail to the lowest-energy (small-polaron) luminescence. They also produce the metastably induced absorption and ESR [84]. These features are all in accord with recent observations on the luminescence of some chalcogenide glasses [39, 87–89].

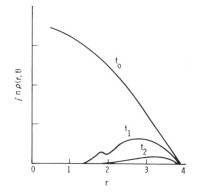

Fig. 2.12 The pair density versus separation for three times after excitation, $t_2 > t_1 > t_0$.

As mentioned previously, a self-trapped charge carrier can be induced to hop by the absorption of a photon of energy roughly equal to the difference between the electronic energies of the initial and final states [46]. In this process, the atoms remain essentially fixed during the excitation process (the Franck–Condon principle). In the absence of electric fields the electronic energy difference is simply due to small-polaron formation. Then the centroid of the absorption band of an isolated small polaron lies at about twice the small-polaron binding energy [46, 84]. However, in the case of oppositely charged small-polaron pairs the centroid of the absorption of each member of the pair is shifted (electromodulated) by the field produced by its partner. As the distribution of pair separations changes in time because of recombination and hopping, the broadening of each small-polaron absorption band is thereby altered. Specifically, as illustrated in Fig. 2.12, the density of small-r pairs, associated with large energy gradients and concomitantly great broadening, falls first (in $> 10^{-9}$ s in a-As$_2$Se$_3$), leaving only the metastably separated (large-r) pairs at long

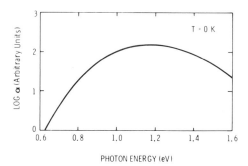

Fig. 2.13 The field-free (unbroadened) hole small-polaron absorption band calculated for a-As$_2$Se$_3$ plotted against photon energy. $E_A = 0.3$ eV; $h\nu = 0.03$ eV; $\alpha/n_p = 10^{-16}$ cm^2.

times (~ 1 s). The small-polaron absorption bands thus narrow in time toward the field-free values of well-separated pairs. As shown in Fig. 2.13, the predicted small-polaron absorption band, with the hole–lattice coupling parameter determined independently from transport experiments on a-As_2Se_3, is similar to the observed metastably induced absorption [82]. Furthermore, consistent with the present work, short-time (picosecond) measurements of the induced absorption in a-As_2S_3 indicate a very broad induced absorption band [89], although the long-time metastably induced absorption band is very much narrower [82]. A further indication of the applicability of the theory would be to produce an increase in the time required for bleaching the induced (midgap) absorption by increasing the energy of the initially absorbed super-band-gap photon.

2.6 SUMMARY AND CONCLUSIONS

The central theme of this chapter is that the transport and optical properties of a semiconductor depend qualitatively on whether or not the charge carriers are best characterized by an itinerant or localized picture. If the carriers find their motion sufficiently impeded, they are then expected to self-trap and form small polarons. They thereby suffer even more extreme localization. The observation of low mobilities and low anomalously signed Hall mobilities in many amorphous semiconductors suggest that the localized view is appropriate in these cases. The measurement of low activated photoconductivity mobilities, geminate recombination, and even super-band-gap luminescence are also in accord with this view. Finally, recent studies of the time dependence of the optically induced midgap absorption and luminescence in chalcogenide glasses are also consistent with this picture.

Nonetheless, much remains to be understood before one can draw conclusions with certainty. In particular, presuming the localized view to be appropriate to many materials, can one develop an understanding of the mechanism of the initial (nonpolaronic) localization? Is the localization of these materials associated with tendencies to form small moleculelike clusters of atoms? If so, can different fabrication techniques alter this situation? For example, might both high- and low-mobility versions of amorphous silicon, dependent on the film morphology, be possible?

In the last decade many views have been reversed. It has now become clear that at least some amorphous semiconductors can be doped, albeit less efficiently than their crystalline counterparts. In addition, the question of the prevalence of defects in what was previously viewed as a defect-free network structure must now be addressed. Here, there can be little doubt that defects play important roles in some situations, such as recombination in amorphous silicon. Although the interaction of the charge carriers with the atomic displacements plays only a perturbative role in many well-studied crystalline semiconductors, in many amorphous semiconductors these interactions are of paramount importance either because the carriers become trapped in deep polaronic traps or because they self-trap to form small polarons.

In considering solar cell applications of amorphous semiconductors, the questions related to localization are central to determining diffusion lengths, recombination lifetimes, and the carrier generation efficiencies. Despite the broad descriptive tone of this article, one must remember that different materials are different. Although the questions which are addressed to understand different materials may be quite similar, the answers may be quite different. Much work must be done before the prospects for solar cell applications of amorphous semiconductors are clear.

ACKNOWLEDGMENT

This work was supported by the U.S. Department of Energy under contract DE-AC04-76-DP00789. Sandia National Laboratories is a U.S. Department of Energy facility.

REFERENCES

1. W. Paul and G. A. N. Connell, *in* "Physics of Structurally Disordered Solids" (S. S. Mitra, ed.), p. 45. Plenum Press, New York, 1976.
2. R. A. Strect and N. F. Mott, *Phys. Rev. Lett.* **35**, 1293 (1975).
3. M. Kastner, D. Adler, and H. Fritzsche, *Phys. Rev. Lett.* **37**, 1504 (1977).
4. R. K. Quinn and R. T. Johnson, Jr., *J. Non-Cryst. Solids* **12**, 213 (1973).
5. N. F. Mott, *Adv. Phys.* **16**, 49 (1976).
6. M. H. Cohen, H. Fritzsche, and S. R. Ovshinsky, *Phys. Rev. Lett.* **22**, 1065 (1969).
7. M. H. Brodsky and R. S. Title, *Phys. Rev. Lett.* **23**, 581 (1969).
8. W. E. Spear and G. H. LeComber, *Solid State Commun.* **17**, 1193 (1975).
9. D. Emin, *in* "Physics of Structurally Disordered Solids" (S. S. Mitra, ed.), p. 385. Plenum Press, New York, 1976.
10. P. W. Anderson, *Phys. Rev.* **109**, 1492 (1958).
11. R. C. Hughes, *Phys. Rev. Lett.* **35**, 449 (1975).
12. E. I. Rashba, *Opt. Spektrosk.* **2**, 75 (1957).
13. Y. Toyozawa, *Prog. Theor. Phys.* **26**, 29 (1961).
14. D. Emin, *Adv. Phys.* **22**, 57 (1973).
15. D. Emin and T. Holstein, *Phys. Rev. Lett.* **36**, 323 (1976).
16. N. F. Mott and A. M. Stoneham, *J. Phys. C* **10**, 3391 (1977).
17. D. Emin, *J. Phys.* (in press).
18. R. C. Hughes and D. Emin, *in* "The Physics of SiO_2 and its Interfaces" (S. Pantelides, ed.), p. 14. Pergamon, Oxford, 1978.
19. D. Emin, *J. Non-Cryst. Solids* **35–36**, 969 (1980).
20. T. Holstein, *Ann. Phys. N.Y.* **8**, 343 (1959).

21. D. V. Lang, R. A. Logan, and M. Jaros, *Phys. Rev. B* **19**, 1015 (1979).
22. P. Pincus, P. Chaikin, and C. F. Coll, III, *Solid State Commun.* **12**, 1265 (1973).
23. S. Lakkis, C. Schlenker, B. K. Chakraverty, R. Buder, and M. Marezio, *Phys. Rev. B* **14**, 1429 (1976).
24. J. Kaplan and C. Kittle, *J. Chem. Phys.* **21**, 1429 (1953).
25. J. C. Thompson, "Electrons in Liquid Ammonia." Oxford University Press (Clarendon), London and New York, 1976.
26. P. W. Anderson, *Phys. Rev. Lett.* **34**, 953 (1975).
27. I. G. Austin and E. S. Garbett, "Electronics and Structural Properties of Amorphous Semiconductors" (P. G. LeComber and J. Mort, eds.), p. 393. Academic Press, New York, 1973.
28. R. C. Hughes, *Phys. Rev.* **15**, 2012 (1977).
29. D. Emin, C. H. Seager, and R. K. Quinn, *Phys. Rev. Lett.* **28**, 813 (1972).
30. D. Emin, *in* "Amorphous and Liquid Semiconductors" (W. E. Spear, ed.), p. 249. Edinburgh Univ. Press, Edinburgh, 1977.
31. A. J. Lewis, *Phys. Rev. B* **14**, 658 (1976).
32. J. Nakahara and K. Kobayaski, *J. Phys. Soc. Jpn.* **40**, 180 (1976).
33. P. K. Ghosh and W. E. Spear, *J. Phys. C* 1347 (1968).
34. J. B. Torrance, M. W. Shafer, and T. R. McGuire, *Phys. Rev. Lett.* **29**, 1168 (1972).
35. D. Emin, *Proc. Int. Conf. Amorphous Liquid Semicond., 8th.*
36. P. M. Pai and R. C. Enck, *Phys. Rev. B* **11**, 5163 (1975).
37. T. D. Moustakas and K. Weiser, *Phys. Rev. B* **12**, 2448 (1975).
38. T. D. Moustakas and W. Paul, *Phys. Rev. B* **16**, 1564 (1977).
39. J. Shah and M. A. Bosch, *Phys. Rev. Lett.* **42**, 1420 (1974).
40. R. C. Zeller and R. O. Pohl, *Phys. Rev. B* **4**, 2029 (1971).
41. P. W. Anderson, B. I. Halperin, and C. M. Varma, *Phil. Mag.* **25**, 1 (1971).
42. D. Emin, C. H. Seager, and R. K. Quinn, *Phys. Rev. Lett.* **28**, 813 (1979).
43. D. Emin, *in* "Electronic and Structural Properties of Amorphous Semiconductors" (P. G. LeComber and J. Mort, eds.), p. 261. Academic Press, New York, 1973.
44. T. Holstein, *Ann. Phys. N.Y.* 343 (1959).
45. D. Emin, *Phys. Rev. Lett.* **32**, 303 (1974).
46. D. Emin, *Adv. Phys.* **24**, 305 (1975).
47. E. Gorham-Bergeron and D. Emin, *Phys. Rev. B* **15**, 3667 (1977).
48. D. Emin, *Phys. Rev. Lett.* **25**, 1751 (1970).
49. D. Emin, *Phys. Rev. B* **3**, 1321 (1971).
50. D. Emin, *Phys. Rev. B* **4**, 3639 (1971).
51. D. Emin and T. Holstein, *Ann. Phys. N.Y.* **53**, 43 (1969).
52. D. Emin, *Solid State Commun.* **22**, 409 (1977).
53. D. Emin, *in* "Linear and Nonlinear Transport in Solids" (J. T. Devreese and V. E. van Doren, eds.), p. 409. Plenum Press, New York, 1976.
54. D. Emin, *Phys. Rev. Lett.* **35**, 882 (1975).
55. C. H. Seager and R. K. Quinn, *J. Non-Cryst. Solids* **17**, 386 (1975).
56. A. J. Lewis, *Phys. Rev. B* **14**, 658 (1976).
57. W. Beyer and H. Mell, *in* "Amorphous and Liquid Semiconductors" (W. E. Spear, ed.), p. 333. Edinburgh Univ. Press, Edinburgh, 1977.
58. E. A. Davis and G. N. Greanes, *in* "Electronic Phenomena in Noncrystalline Semiconductors" (T. T. Kolomietz, ed.), p. 212. Nauka, Leningrad, 1976.
59. T. N. Kennedy and J. D. Mackenzie, *Phys. Chem. Glasses* **8**, 169 (1967).
60. P. Nagels, R. Callaerts, and M. Denayer, *in* "Amorphous and Liquid Semiconductors" (J. Stuke and W. Brenig, eds.), p. 867. Taylor and Francis, London, 1974.
61. N. F. Mott, E. A. Davis, and R. A. Street, *Phil. Mag.* **32**, 961 (1975).

62. E. Mytilineau and E. A. Davis, *in* "Amorphous and Liquid Semiconductors" (W. E. Spear, ed.), p. 632. Edinburgh Univ. Press, Edinburgh, 1977.

63. G. H. Döhler, *Phys. Rev. B* **19**, 2083 (1979).

64. T. Holstein, *Phys. Rev.* **124**, 1329 (1961).

65. D. Emin and T. Holstein, *Ann. Phys. N.Y.* **53**, 439 (1969).

66. D. Emin, *Phil. Mag.* **35**, 1188 (1977).

67. N. F. Mott, *Phil. Mag.* **38**, 549 (1978).

68. K. W. Boer, *Phys. State Solidi* **34**, 721 (1969).

69. E. J. Yoffa and D. Adler, *Phys. Rev. B* **15**, 2311 (1977).

70. H. J. deWit, *J. Appl. Phys.* **43**, 908 (1972).

71. P. Nagels, R. Callaerts, M. Denayer, and R. DeConink, *J. Non-Cryst. Solids* **4**, 295 (1970).

72. D. Meimaris, J. Katris, D. Martakos, and M. Roilos, *Phil. Mag.* **35**, 1633 (1977).

73. A. H. Clark, *Phys. Rev.* **154**, 750 (1967).

74. A. Miller and E. Abrahams, *Phys. Rev.* **120**, 745 (1960).

75. G. N. Greaves, *J. Non-Cryst. Solids* **11**, 427 (1973).

76. N. F. Mott, *Phil. Mag.* **19**, 835 (1969).

77. D. Emin, *in* "Physics of Structural Disordered Solids" (S. S. Mitra, ed.), p. 461. Plenum Press, New York, 1976.

78. M. H. Brodsky and R. J. Gambino, *J. Non-Cryst. Solids* **8–10**, 439 (1972).

79. M. L. Knotek, M. Pollak, and T. M. Donovan, *Phys. Rev. Lett.* **30**, 853 (1973).

80. F. Gutmann and L. E. Lyons, "Organic Semiconductors," p. 296. Wiley, New York, 1967.

81. P. O'Connor and J. Tauc, *Phys. Rev. Lett.* **43**, 311 (1979).

82. S. G. Bishop, U. Strom, and P. C. Taylor, *Phys. Rev. B* **15**, 2278 (1977).

83. M. Klinger, *Phys. Lett.* **1**, 102 (1963).

84. D. Emin, *in* "Amorphous and Liquid Semiconductors" (W. E. Spear, ed.), p. 261. Edinburgh Univ. Press, Edinburgh, 1977.

85. C. Benoit a la Guillaume, F. Mollot, and J. Cernogora, *in* "Amorphous and Liquid Semiconductors" (W. B. Spear, ed.), p. 612. Edinburgh Univ. Press, Edinburgh, 1977.

86. R. A. Street, *Adv. Phys.* **25**, 397 (1976).

87. M. A. Bosch and J. Shah, *Phys. Rev. Lett.* **42**, 118 (1979).

88. M. A. Bosch, R. W. Epworth, and D. Emin, *J. Non-Cryst. Solids* (in press).

89. R. L. Fork, C. V. Shank, A. M. Glass, A. Migus, M. A. Bosch, and J. Shah, *Phys. Rev. Lett.* **43**, 394 (1979).

3 Electrical Properties of Polycrystalline Semiconductor Thin Films

LAWRENCE L. KAZMERSKI

Photovoltaics Branch
Solar Energy Research Institute
Golden, Colorado

3.1 INTRODUCTION

This chapter focuses on the basic electronic transport in polycrystalline *semiconductor* thin films. A number of books [1] and reviews are available that cover in detail the properties of discontinuous [2], metal [3, 4], and insulator films [5]. Except for earlier reviews by Anderson [6] and Bube [7], the treatment of carrier transport mechanisms in polycrystalline semiconductor thin films has been largely heuristic and superficial. Although

59

identification of the electrical properties of these films is an expanding area of research, many of the major contributions to this field remain segmented in the literature. It is the purpose of this chapter to delineate the basic mechanisms involved, to integrate and compare some recent contributions with earlier work, and to provide a general basis for understanding semiconductor thin-film properties.

In this chapter, carrier transport is considered separately for two cases. First, the electrical characteristics of perfect, single-crystal thin films are examined. These transport properties are regarded as essentially those of the bulk crystalline semiconductor, but they are altered by the major physical feature of the film—the surface. Flat-band and surface-band bending conditions are investigated, and film thickness effects are identified. Second, transport mechanisms in polycrystalline films are discussed. In this case, the complicating factors of film defects and discontinuities are considered, with some emphasis on the grain boundary. The combined effects of surface scattering and defect-dominated properties are also indicated.

At the onset of this chapter, it must be emphasized that the identification and definition of electronic transport in polycrystalline semiconductor films has largely been a *modeling effort*. At this writing, no universal explanation of transport characteristics exists for thin films. Differences should be anticipated. For example, polycrystalline compound semiconductor films can gain their extrinsic character by stoichiometry control, but elemental semiconductors by doping. Grain boundaries differ not only between these two types but also between large-grained and small-grained materials. Understanding and control of the electrical properties of thin semiconductor films has now progressed from a scientific curiosity to a necessity. The potential economical large-scale deployment of devices based upon these thin films depends upon substantial progress in this area. While the thin-film research veteran may express caution or hesitancy because of the magnitude or seeming unsolvability of the problems, the challenge envisioned by others may bring about the solutions. It is hoped that this chapter might serve especially the latter group.

3.2 TRANSPORT IN THIN CRYSTALLINE FILMS

1. Definitions and Formulations

In this section, the most simple case of a semiconductor thin film is considered—the continuous single-crystal film. Conceptually, the situation avoids the complicating factors of film defects, defect structures (e.g., dislocations, stacking faults, and grain boundaries), and other discontinuities that may affect conduction mechanisms. Therefore, the transport properties can be regarded as essentially those of the bulk crystalline material, influenced and altered by the remaining major physical feature of the film—the surface.

The surface of a thin film affects the electrical transport properties of a material, whether a semiconductor or a metal, by limiting the traversal of the charge carriers and their mean free paths. Even bulk materials experience these surface effects, but they are more pronounced in thin films because of the large surface-to-volume ratios. When the thickness of the film becomes less than or comparable to the mean free path of the carriers, the scattering of the electrons and holes from the film surfaces has measurable effects on carrier transport properties and can dominate the electrical characteristics of the film. Of course, the extent of the influence of surface scattering depends upon the nature of the scattering mechanism(s) involved. The two limiting cases are

(1) *Specular reflection*: During the scattering process, the carriers (electrons or holes) have only their velocity component perpendicular to the surface *reversed*, and their energy remains constant. Since no losses occur, there is no effect on conductivity. The surface represents a perfect reflector, and the scattering is elastic.

(2) *Diffuse reflection*: After scattering, the carriers emerge from the surface with velocities independent of their incident ones. This process is indicative of inelastic or random scattering. The change in momentum leads to a related change in conductivity.

Specular reflection is the type of scattering expected from an *ideal* surface. *Real* surfaces exhibit some amount of disorder which, in turn, results in some degree of diffuse scattering. The extent of diffuse scattering is determined by the type, density, and cross sections of the surface defects. The major mechanisms, primarily surface charge impurities and electron–phonon interactions, that determine the extent of diffuse or specular scattering have been treated in detail by Greene [8–10] and Tavger *et al.* [11–13]. Quantitatively, total diffuse scattering can lower effective film conductivities and mobilities more than an order of magnitude below their single-crystal bulk magnitudes. For a real surface, both partial specular and partial diffuse scattering mechanisms exist, and the resultant electrical properties usually lie between those predicted by either scattering mode.

This section discusses the perfect semiconductor thin film, relating the effective film transport properties as influenced by surface conditions to the predicted bulk behavior. Bulk properties are summarized, and the

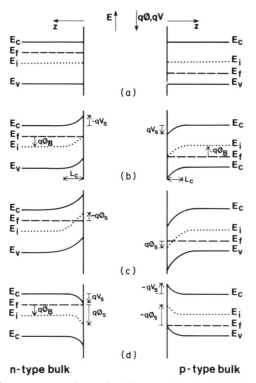

Fig. 3.1 Band structure near the semiconductor surface for (a) flat-band, (b) depletion, (c) inversion, and (d) accumulation conditions. Both *n*- and *p*-type bulk semiconductors are represented.

effects of diffuse and specular reflection are related. Two surface conditions are considered.

The first models the film with energy bands constant to the surface itself. This is commonly called the *flat-band condition* and is illustrated schematically in Fig. 3.1a. Although this case is somewhat artificial and does not exactly represent the situation for the real semiconductor surface region, it is instructive and provides a basis for understanding the various scattering phenomena and their effects. Because of its simplicity, the flat-band model can be used to predict transport properties and has been used to explain the general electrical behavior of some films.

The second model includes *band-bending* at the surface. The existence of surface states complicates the transport mechanisms. Band structures for surface depletion, inversion, and accumulation are illustrated in Fig. 3.1b–d. Effects of band bending on semiconductor film properties have been studied, but it is not easy to control surface state densities and

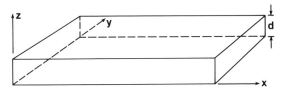

Fig. 3.2 Thin-film geometry and coordinate system.

associated surface potentials by or during film growth processes. Field-effect techniques, discussed later in this section, can be utilized to demonstrate the relationships between the degree of the non-flat-band condition and the resulting electrical properties of the thin semiconductor films.

For all film geometries in this chapter, the coordinate system shown in Fig. 3.2 has been adopted. The z direction is perpendicular to the film surface, and the total film thickness defined in this direction is d.

2. Essential Transport Phenomena

In evaluating electrical characteristics of thin films, it is common to compare the film's behavior to that of the bulk crystal. If the bulk crystal were perfect, the electrons and holes could flow unimpeded in a perfect periodic potential. In the absence of external fields, each carrier would maintain its velocity and wave vector indefinitely. However, in the real bulk crystal lattice, vibrations, impurities, and defects could cause deviations from the ideal behavior. The carriers experience a nonzero scattering probability which provides for random movement and continual velocity change for the electrons or holes. A drift current results when an electric field is imposed.

The various scattering processes can each be characterized by a fundamental relaxation time τ defined as the average time required for a disturbance in the electron distribution to fade by the random action of the scattering. Two scattering processes are especially important for bulk crystal behavior: first, scattering by lattice vibrations, dominant in chemically pure crystals at ordinary temperatures, and second, scattering from impurity centers. For the first case, the relaxation time is [14]

$$\tau_l \sim T^{-1} E^{-1/2} \tag{3.1}$$

where T is the temperature and E the carrier energy. In case 2, scattering occurs from impurity centers, such as ionized impurities. In this case [14]

$$\tau_i \sim E^{3/2} / N_i \tag{3.2}$$

where N_i is the density of ionized impurities. All scattering mechanisms act simultaneously to some extent in real crystals. The total bulk relaxation time, considering each of the scattering mechanisms to be essentially independent, is given by [14]

$$1/\tau_b = 1/\tau_l + 1/\tau_i + \cdots \tag{3.3}$$

In order to relate the effects of the various scattering phenomena to determination of the transport equations, the relaxation times are incorporated into the Boltzmann transport equation [15]. From this, the current–field relationship can be generated and the pertinent transport parameters (mobility, conductivity, carrier concentration) can be extracted. Following this procedure, the temperature dependencies of the lattice and the ionized impurity scattering carrier mobilities can be expressed [14]

$$\mu_l = CT^{-3/2} \tag{3.4}$$

and

$$\mu_i = C'T^{3/2} \tag{3.5}$$

respectively, where C and C' are constants. In addition, when both types of scattering are present and noninteracting, the mobilities "add" according to the relaxation times [Eq. (3.3)], with the effective bulk crystalline mobility given by [16].

$$1/\mu_b = 1/\mu_l + 1/\mu_i \tag{3.6}$$

From Eqs. (3.4) and (3.5) it is apparent that lattice scattering predominates at high temperatures and impurity scattering at lower temperatures. The transition from domination by one type to the other depends upon the material and the impurity nature and concentration. Some deviations from the "3/2 law" are encountered in real semiconductors due to complex band structures as well as optical phonon scattering. Experimentally it has been found that the mobility variations usually range from $T^{-5/2}$ to $T^{5/2}$ dependencies. For example, near room temperature the electron and hole mobilities for GaP follow Eq. (3.4), whereas for silicon, $\mu_n \sim T^{-5/2}$ [17].*

3. The Effects of Surfaces on Carrier Transport

The bulk transport case summarized briefly in the previous section must be modified even for the single-crystal thin film, since the surface can scatter a greater number of carriers than the number not being scattered. It

---

*In general, the temperature dependence of the mobility of a crystalline semiconductor can be expressed [13] as

$$\mu \sim T^{\alpha + \beta/2} \tag{3.7}$$

where α and β are constants whose magnitudes and signs are indicative of the dominant scattering mechanism(s).

is assumed that the flat-band condition holds for the analysis in this section. The perturbations due to band bending are included in Section 3.4.

A. FLAT-BAND CONDITIONS: RIGOROUS TREATMENT

The method utilized to incorporate the effects of surface scattering is similar to that cited for bulk phenomena in the previous section and starts with the Boltzmann equation [15]:

$$a \, \nabla_c f + c \, \nabla_r f = (\partial f / \partial t)_{\text{scattering}}$$ (3.8)

where a and c are the acceleration and the velocity of the scattered carriers, respectively, and f the distribution function. This theoretical approach applied to surface scattering has been reported by Fuchs [18], Sondheimer [19], Lucas [20], Tavger [13], Zemel [21], Anderson [6], Frankl [22], and Fleitner [23]. Reviews have also been compiled by Campbell [24], Mayer [25], Chopra [26], and Many *et al.* [27]. When an electric field $\mathcal{E}$ is applied in the x direction, Eq. (3.8) reduces to a one-dimensional representation for the geometry shown in Fig. 3.2, in which the x and y film dimensions are very much greater than d, the film thickness. Therefore,

$$\frac{-q\mathcal{E}_x}{m^*} \frac{\partial f}{\partial c_x} + c_z \frac{\partial f}{\partial z} = \frac{f - f_0}{\tau}$$ (3.9)

The distribution function can be written [6, 27] as

$$f = f_0 + f_1(c, z)$$ (3.10)

since the perturbation f_1 is independent of x and y. The general form of the solution of the differential equation [Eq. (3.9)] is [27]

$$f_1 = \frac{q\mathcal{E}_x \tau}{m^*} \frac{\partial f_0}{\partial c_x} \left[1 + F(c) \exp\left(\frac{-z}{\tau c_z} \right) \right]$$ (3.11)

where the functional form of $F(c)$ depends upon the boundary conditions.

If the upper and lower surfaces of the film are identical, from the symmetry of the situation,

$$f_1(c_{x, y}, c_z; z) = f_1(c_{x, y}, -c_z; d - z)$$ (3.12)

Therefore, two solutions of Eq. (3.12) are found [6, 27]:

$$f_1 = f_1^+ = \frac{q\mathcal{E}\tau}{m^*} \frac{\partial f_0}{\partial c_x} \left[1 + F(c) \exp\left(\frac{-x}{\tau c_z} \right) \right], \qquad c_z \geqslant 0$$ (3.13a)

and

$$f_1 = f_1^- = \frac{q\mathcal{E}_\tau}{m^*} \frac{\partial f_0}{\partial c_x} \left[1 + F(-c) \exp\left(\frac{d - z}{\tau c_z} \right) \right], \qquad c_z \leqslant 0$$ (3.13b)

However, it has already been noted that carriers can be scattered diffusely from the surface, and the solutions given by Eqs. (3.13a) and (3.13b) represent the specular case only. The condition of the surface and its effect on the carrier transport can now be introduced in a simple fashion. From Eq. (3.10) the distribution function of carriers *arriving* at the surface is given by

$$f' = f_0 + f_1^+ \qquad (3.14)$$

If some fraction p of these undergoes *specular reflection*, their resultant distribution function *leaving* the surface is

$$f'' = p(f_0 + f_1^-) \qquad (3.15)$$

The fraction p is called the specular scattering factor, with $p = 1$ indicating pure specular reflection and $p = 0$ entirely diffuse reflection. The remaining carriers leaving the surface, $1 - p$, are scattered diffusely, and their distribution function is

$$f''' = (1 - p)f_0 \qquad (3.16)$$

But

$$f' = f'' + f''' \qquad (3.17)$$

or

$$f_0 + f_1^+ = p(f_0 + f_1^-) + (1 - p)f_0 \qquad (3.18)$$

Substituting Eqs. (3.13a) and (3.13b) into Eq. (3.18) yields [6]

$$F(c) = -(1 - p)/[1 - p\exp(-d/\tau c_z)] \qquad (3.19)$$

The carrier current density is calculated by inserting $F(c)$ into Eq. (3.11) and integrating the product of the velocity, density of states, and distribution functions; that is,

$$J_x = -q\int_c c_x N_c f_1\,dc = nq\mu' \qquad (3.20)$$

where μ' is the effective film mobility.

Two solutions for the mobility result, depending on whether the semiconductor is degenerate or nondegenerate. For the former case, in which Firmi–Dirac statistics apply, the ratio of the effective to the bulk mobility is [6, 27]

$$\frac{\mu'}{\mu_b} = 1 - \frac{3(1-p)}{2(d/\lambda)}\int_1^\infty \left(\frac{1}{\xi^3} - \frac{1}{\xi^5}\right)\left\{\frac{1 - \exp[-(d/\lambda)\xi]}{1 - p\exp[-(d/\lambda)\xi]}\right\}d\xi \qquad (3.21)$$

where λ is the carrier mean free path. For the nondegenerate semiconductor case having spherical energy surfaces, Boltzmann statistics apply, and [6]

$$\mu'/\mu_b = 1 - (1 - p)(2\lambda/d) + (1 - 2p)(2\lambda/d)\Gamma_1(2\lambda/d)$$
$$+ (2p\lambda/d)\Gamma_1(\lambda/d) \qquad (3.22)$$

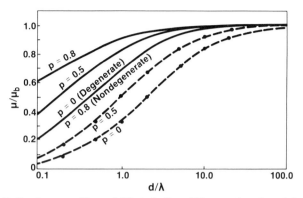

Fig. 3.3 Ratio of average film mobility to bulk mobility as a function of the ratio of film thickness to mean surface scattering length using both Fermi statistics [solid lines, representing the degenerate semiconductor case of Eq. (3.21)] and Boltzmann statistics [broken lines, representing the nondegenerate semiconductor case of Eq. (3.22)]. The effects of specular and diffuse scattering are shown by the difference in the specularity factor p. The solid circles represent the approximate solution given by Eq. (3.29) (from Anderson [6]).

where

$$\Gamma_1(n\lambda/d) = \int_0^\infty \exp\left[-\epsilon - (d/n\lambda)(\pi\epsilon)^{-1/2}\right] d\epsilon \qquad (3.23)$$

and $n = 1$ or 2.

The functional dependencies of the mobility ratios for the degenerate and nondegenerate cases [predicted by Eqs. (3.21) and (3.22), respectively] are shown in Fig. 3.3. Surface scattering is more dominant for the nondegenerate semiconductor than for the corresponding degenerate semiconductor or metal. As the semiconductor becomes more degenerate, the effect of surface scattering becomes less significant.

Anisotropy of the effective mobility can result if the semiconductor has nonspherical equal-energy surfaces. The case of ellipsoidal energy surfaces has been considered by Ham and Mattis [28]. Their refinement of the previous spherical case indicates a fractional difference in mobility magnitudes for different crystal directions. Figure 3.4 presents their results for the completely diffuse ($p = 0$) scattering case, for a diamond structure nondegenerate semiconductor. The symbol n refers to the direction normal to the film surface, and j is the current density. For a given thickness, the value of μ' depends significantly upon the current direction and can be as much as an order of magnitude lower than that presented in Fig. 3.3. Thus the semiconductor thin film might be even more sensitive to the surface condition and film thickness than the previous analysis indicates, and some caution must be exhibited in applying this interpretation for transport in thinner semiconductor films.

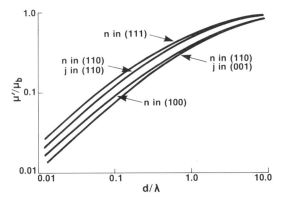

Fig. 3.4 Ratio of effective film mobility to bulk mobility as a function of the ratio of film thickness to mean surface scattering length for the case of ellipsoidal energy surfaces, indicating the anisotropy of the electrical properties (from Ham and Mattis [28]).

B.   FLAT-BAND CONDITIONS: SIMPLIFIED APPROACH

Another more simple approach can be used to predict the effect of surface scattering and film thickness on transport properties. For a nondegenerate semiconductor, assuming flat bands and noninteracting scattering mechanisms, the total relaxation time using Matteissen's rule [16] is

$$1/\tau' = 1/\tau_s + 1/\tau_b \tag{3.24}$$

where τ_s represents the average time a carrier requires to collide with the surface toward which it is moving and τ_b is defined in Eq. (3.3). The mean free distance of a carrier from a surface is approximately the film half-thickness $d/2$. If the unilateral mean velocity v_z is defined as the average over the positive (or negative) z-direction velocity component of all carriers, then the average surface scattering time can be estimated by [27]

$$\tau_s \simeq (d/2)/v_z \tag{3.25}$$

But since the unilateral mean free path λ is defined as [27]

$$\lambda = \tau_b v_z \tag{3.26}$$

Then

$$\tau_s \simeq (d/2\lambda)\tau_b \tag{3.27}$$

Combining Eqs. (3.24) and (3.27), the effective mobility becomes [27]

$$\mu' = \mu_b/(1 + 2\lambda/d) \tag{3.28}$$

This simple derivation assumes that the surface scattering is entirely specular ($p = 1$). To generalize to the case in which a fraction p of the carriers is scattered inelastically (diffuse scattering), it is necessary to adjust the expression for τ_s, which represents the reciprocal per unit time probability that an electron will be scattered by a surface. Since p is the fraction undergoing specular (elastic) scattering, $1 - p$ is the fraction undergoing diffuse (inelastic) scattering, and the surface scattering term in Eq. (3.28) becomes $(1 - p)/\tau_s$. Thus, the mobility can be written [6, 27]

$$\mu' = \mu_b\left[1 + (1 - p)(2\lambda/d)\right]^{-1} \tag{3.29}$$

The good agreement of Eq. (3.29) with the more rigorously derived Eq. (3.22) is shown in Fig. 3.3.

The conditions of the upper and lower surfaces of a thin film might be expected to be quite different, since one is in contact with a supporting substrate and the other is exposed to a dissimilar environment (gas, solid, or liquid). In this situation, the contributions from each surface can be averaged (a first-order approximation), and p can be replaced by $(p + q)/2$, where p and q are the specular scattering coefficients from the upper and lower film surfaces, respectively. Thus, Eq. (3.29) becomes [29]

$$\mu' = \mu_b\left\{1 + (2\lambda/d)\left[1 - \tfrac{1}{2}(p + q)\right]\right\}^{-1} \tag{3.30}$$

This same substitution can be incorporated directly into Eq. (3.22), keeping the formulation and the results consistent.

C. EFFECTIVE SURFACE SCATTERING LENGTH

In effect, the parameter λ is a measure of the influence of the surface upon the carrier transport. The surface scattering length is defined in terms of the mean velocity of the carriers [27, 29, 30]:

$$\lambda = v_z \tau_b \tag{3.31}$$

where τ_b is the bulk relaxation time. The velocity component is measured perpendicular to the film surface over which it is being averaged. This mean velocity is given by

$$v_z = (kT/2m^*)^{1/2} \tag{3.32}$$

Therefore, combining Eq. (3.32) with Eq. (3.31) and noting that $\mu_b = q\tau_b/m^*$, one can express the mean surface scattering length as [30]

$$\lambda = C'\mu_b T^{1/2} \tag{3.33}$$

where $C' = (m^*k/2\pi q^2)^{1/2}$. Figure 3.5 shows the dependence of λ/μ_b on temperature, indicating the predicted dependence of Eq. (3.33) for CdS films [29].

The magnitude of λ is material dependent. For carrier concentrations comparable to CdS (i.e., $10^{15}-10^{16}/cm^3$), $\lambda = 1.95, 0.85$, and $0.12\ \mu m$ for Si, GaAs, and Ge, respectively [31]. Thus, surface scattering is expected to affect the electrical properties of Si more than those of the other semiconductors at a given film thickness. For a 1-μm-thick film, the mobility of Ge

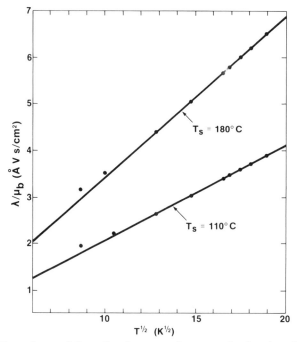

Fig. 3.5 Dependence of the ratio of mean surface scattering length to bulk mobility upon film temperature, for substrate temperatures $T_s = 110°C$ (383 K) and 180°C (453 K), indicating a $T^{1/2}$ relationship for CdS thin films (from Kazmerski [30]).

is expected to be only about 5% less than its bulk value, while that for Si is about 55% less.

4.   Effects of Band Bending at the Surface

The derivations of the previous sections assumed that the semiconductor energy bands were constant from the bulk to the surface itself. If surface states exist, some degree of band bending at the surface will result under the following conditions:

(1) *Depletion*: Majority carriers can be trapped in acceptor-type surface states (n-type semiconductor) or donor-type surface states (p-type semiconductor) at energy levels below or above the Fermi level, respectively. In the case of an n-type semiconductor, electrons are repelled from the surface region, making it less n-type (i.e., the surface is depleted of electrons). The band edges will *bend up* at the surface, away from the Fermi level. If this bending is sufficient, the surface can become p-type (i.e., the Fermi level is closer to the valence band), and this condition is called *inversion*. For a p-type semiconductor, holes are repelled from the

surface and the band edges *bend down* toward the Fermi level. Once again, if this bending is sufficient, the surface region can be inverted (i.e., become *n*-type). The effect is equivalent to applying a *negative bias* to the surface of the *p*-type semiconductor, or a *positive bias* to the surface of the *n*-type semiconductor.

(2) *Accumulation*: This situation is the converse of depletion. Donor states at the surface of an *n*-type semiconductor, or acceptor states for a *p*-type semiconductor, contribute additional majority carriers (accumulation of majority carriers) to the conduction or valence band, respectively. The band edges will bend toward the Fermi level (i.e., bend downward for the *n*-type and upward for the *p*-type semiconductor). The effect here is the application of a *positive* bias to the surface of the *p*-type semiconductor, or a *negative bias* to the surface of the *n*-type semiconductor.

Depletion and accumulation are illustrated in Fig. 3.1 in comparison to the flat-band case. Generally, the free surface of an impurity semiconductor is in *depletion* unless external fields are applied. The depletion condition will therefore receive emphasis in subsequent sections.

A. THE SURFACE SPACE CHARGE REGION

Figure 3.1 shows that the extent of the penetration of the surface depletion region is significant. The width of this region L_c depends upon the condition of the surface (surface charge, surface potential) and the condition of the bulk semiconductor (doping concentrations, Fermi level position, intrinsic concentration). For example, Waxman *et al.* [32] calculated the effective layer thickness for CdS as a function of the surface potential $(v_s = qV_s/kT)$, as shown in Fig. 3.6. The change in L_c is predicted to be more than an order of magnitude for a corresponding 0–0.2-eV change in qV_s. Many *et al.* [27] carefully considered this problem, using a solution of the Poisson equation

$$d^2v/dz^2 = -\rho/\epsilon kT = -(q^2/\epsilon kT)\big[n_b - p_b + p_b\exp(-v) - n_b\exp(v)\big]$$

(3.34)

where ϵ is the semiconductor permittivity, n_b and p_b the bulk carrier concentrations, and v a dimensionless potential $(= qV/kT)$. The potential barrier V is defined as the potential at any point in the space charge region with respect to the value in the bulk (i.e., $V = \phi - \phi_b$).

For the case of small perturbations (e.g., $|v| \lesssim \frac{1}{2}$, the integration of Eq. (3.34), applying the boundary condition $[(dv)/(dz)]|_{z=0} = 0$, yields [26]

$$dv/dz = \mp F(v, u_b)/L$$

(3.35)

where L is the effective Debye length $(= [\epsilon kT/q^2(n_b + p_b)]^{1/2})$, u_b a

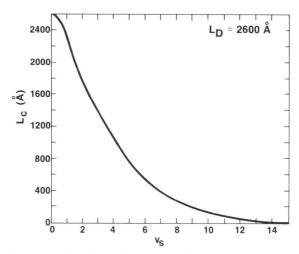

Fig. 3.6 Calculated surface layer thickness as a function of surface potential for a CdS film with a bulk concentration of $10^{14}/cm^3$. L_D is the Debye length (from Waxman *et al.* [32]).

reduced potential ($= q\phi_b/kT$), and [26]

$$F(v, u_b) = 2^{1/2}\left[\frac{\cosh(u_b + v)}{\cosh u_b} - v\tanh u_b - 1\right]^{1/2} \qquad (3.36)$$

The minus sign in Eq. (3.35) refers to $v > 0$, and the plus sign to $v < 0$.

The resultant potential penetration profile was obtained by Many *et al.* by numerically integrating the complex Eq. (3.35). The potential barrier $|v|$ is shown as a function of z/L, a normalized distance from the surface (Fig. 3.7). This profile is significant for thin films. The band diagram for the depletion condition (Fig. 3.1b) shows that the reduced potentials v and u_b are opposite in sign. The flat-band condition exists when $v(0) = 0$. Band bending continues until $v(0) = -2u_b$. In this condition, the semiconductor is quasi-intrinsic, since the minority carrier density equals the majority carrier density in the bulk (i.e., in terms of the band diagram, the Fermi level now lies *below* E_i, the midgap energy, by the same amount u_b that it was positioned *above* it in the flat-band condition). As the band bending continues [$|v(0)| > 2u_b$], the total inversion condition is reached and the surface undergoes a change in majority carrier type.

Consider a film with $|u_b| = 6$ (i.e., the Fermi level lies about 0.15 eV above midgap for the *n*-type semiconductor at room temperature). At $z/L \sim 0.7$, the quasi-intrinsic condition holds. Therefore, if $d \sim 2(z/L)$ or approximately 1.4, the film will appear intrinsic. Anderson observes that this situation corresponds physically to having the total number of carriers

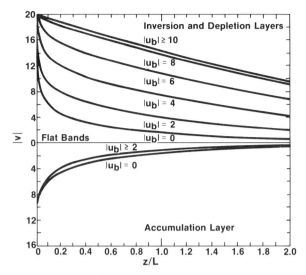

Fig. 3.7 The shape of the barrier potential as a function of the normalized distance from the surface (L is the Debye length) for various values of the bulk potential. The magnitude of $|v_s|$ is taken to be 20 (from Many *et al.* [27]).

dominated by the surface trapping mechanism because of the small volume of the film. It can be observed further that, for higher doping levels ($|u_b| > 6$), quite thinner films would be needed to provide the intrinsic appearance.

An interesting situation results when the band bending occurs at both the upper and lower surfaces of the film. For a sufficiently thin film, the conduction and valence band edges cannot reach the positions with respect to the Fermi level that would be expected for a corresponding bulk material with the pertinent doping level. This situation is illustrated simplistically in Fig. 3.8. It is interesting to note that, if the film is thin enough, the band bending can develop only to a small extent [33, 34]. As a result, the potential and carrier concentrations appear almost uniform throughout the film, and the transport properties are characterized by the less complicated flat-band models [e.g., Eqs. (3.22) and (3.29)].

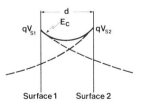

Fig. 3.8 Effective band structure resulting from band bending at surfaces in a very thin semiconductor film.

B. SURFACE TRANSPORT PARAMETERS

A major effect of the band-bending phenomenon is the generation of an excess or a deficiency of mobile carriers within the surface region. These quantities are defined [27] as

$$\Delta N = \int_0^\infty (n - n_b)\, dz \tag{3.37a}$$

and

$$\Delta P \int_0^\infty (p - p_b)\, dz \tag{3.37b}$$

where $p = p_b \exp(-v)$ and $n = n_b \exp(v)$. Since ΔN and ΔP are surface quantities, they have per-unit-area units. Many et al. [27] solved these integrals numerically for accumulation, depletion, and inversion conditions. These parameters can be related to the carrier transport, since ΔN and ΔP effect a change in the surface conductivity:

$$\Delta\sigma = q\left(\mu_{ns}\,\Delta N + \mu_{ps}\,\Delta P\right) \tag{3.38}$$

where $\Delta\sigma$ is expressed per unit surface area and μ_{ns} and μ_{ps} are the electron and hole surface mobilities, respectively. The surface conductivity change can be measured, although it depends greatly on the nature and magnitude of the surface potential. Although it is fairly simple to measure the surface conductance, it is impossible to separate the product $\mu_{ns}\,\Delta N$ (or $\mu_{ps}\,\Delta P$) without some further theoretical estimates.

The surface mobilities μ_{ns} and μ_{ps} can be calculated, and their relationships to the surface conditions can be predicted. The general case of nonparabolic bands for a nondegenerate or degenerate semiconductor has been treated by Juhasz [35]. For either the depletion or accumulation he showed that

$$\mu_{ns}/\mu_b = 1 - (\lambda n_b/\Delta N)H_n(v) \tag{3.39}$$

and

$$\mu_{ns}/\mu_b = 1 - (\lambda p_b/\Delta P)H_p(v) \tag{3.40}$$

where n_p and p_b are bulk concentrations and $H_n(v)$ and $H_p(v)$ are functions which reduce to the Γ functions derived by Many et al. [27] and are presented in Eq. (3.23) for a nondegenerate semiconductor. The dependence of μ_{ns}/μ_b on surface potential is presented in Fig. 3.9 for both degenerate and nondegenerate cases. For the nondegenerate case, μ_{ns}/μ_b decreases with increasing v_s as expected. However, a maximum of unity (i.e., $\mu_{ns} = \mu_b$) is not predicted for the $v_s = 0$ (i.e., no band bending) case. In this limit, μ_{ns} corresponds to the surface mobility with normal diffuse scattering. For the degenerate film, the reasons for the mobility cusp as v_s

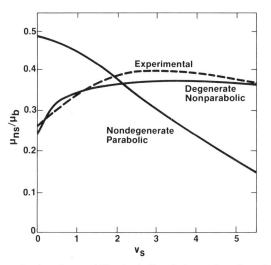

Fig. 3.9 Normalized surface mobility (to bulk value) as a function of surface potential for degenerate and nondegenerate cases. Data for degenerate InSb thin film are included (from Juhasz [35]).

approaches zero have been discussed by Greene [9] and Frankl [22]. Similar electrical characteristics have been derived by Covington and Ray [33] and Hezel [36].

Anderson [6] calculated the effective mobilities for the band-bending case by modeling the surface region as shown in Fig. 3.10. The surface region of width L_c is approximated by two independent mean scattering times. The first is associated with *bound carriers*. These carriers are constrained to move in the surface potential well and scatter diffusely at the surface ($z = 0$) but specularly at the boundary ($z = L_c$). The second is associated with *unbound electrons*. These have energies above the well and

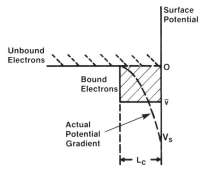

Fig. 3.10 Potential well model for calculating scattering at a surface (from Anderson [6]).

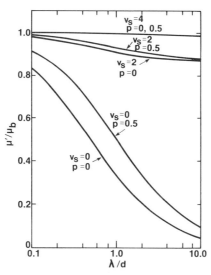

Fig. 3.11 Dependence of the ratio of film to bulk mobility upon the ratio of mean surface scattering length to film thickness for various surface potentials ($v_s = qV_s/kT$), using the nondegenerate model given by Eq. (3.44). Effects of specular and diffuse scattering are also indicated.

are scattered at the surface under a flat-band condition modified by the surface potential. The resultant mean surface scattering times are given by

$$1/\tau_s = 1/\tau_s(\text{bound}) + 1/\tau_s(\text{unbound}) \tag{3.41}$$

$$\frac{1}{\tau_s} = \frac{(1-p)\lambda(1+v_s)^{1/2}}{L_c\tau_b} + \frac{(1-p)\lambda(1+v_s)^{1/2}}{d\tau_b} \tag{3.42}$$

When this modified surface scattering time is substituted into Eq. (3.24) for a surface accumulation layer, the effective mobility is [6]

$$\mu'/\mu_b = \left[1 + (1-p)\lambda(1/L_c + 2/d)(1 + qV_s/kT)^{1/2}\right]^{-1} \tag{3.43}$$

For the more common depletion layer at the surface of the semiconductor film [6],

$$\frac{\mu'}{\mu_b} = \frac{1 + (1-p)(2\lambda/d)\left[1 - \exp(qV_s/kT)\right]}{1 + (1-p)(2\lambda/d)} \tag{3.44}$$

Equations (3.43) and (3.44) are both consistent with the expression derived for the flat-band situation. As $V_s \to 0$ and $L_c \to \infty$, these equations are identical to Eq. (3.29). The effects of the surface potential on the mobility are illustrated in Fig. 3.11.

C. HALL-EFFECT PARAMETERS

The previously derived expressions for film mobilities apply to the measurement of Hall mobility under small (normal) magnetic fields. However, the measured carrier concentrations are surface scattering- and thickness-dependent, just as for the case of mobility or resistivity, when the number of carriers being scattered from the surface is much greater than the number not being scattered. The relationship of this scattering mechanism to the magnitude of the Hall constant has been treated by Amith [37] who solved the Boltzmann equation for an extrinsic semiconductor with an additional drift field due to an applied magnetic field. The result indicates that the effective Hall constant R_H' is related to the crystalline quantity R_{Hb} through the expression

$$R_H' = \eta(\lambda/d)R_{Hb} \tag{3.45}$$

where the solution of the Boltzmann equation yields

$$\eta\left(\frac{\lambda}{d}\right) = \frac{1 - 4\lambda/d + 4(\lambda/d)\Gamma_1(\lambda/d) + \Gamma_3(\lambda/d)}{\left[1 - 2\lambda/d + (2\lambda/d)\Gamma_1(\lambda/d)\right]^2} \tag{3.46}$$

and $\Gamma_3 = \int_0^\infty (\epsilon\pi)^{-1/2}\Gamma_1(\lambda/d)\,d\epsilon$ with $\Gamma_1(\lambda/d)$ defined in Eq. (3.23). Values for Γ_1 and Γ_3 can be obtained by numerical integration, and the resulting general dependence of $\eta(\lambda/d)$ upon the mean scattering length and thickness is shown in Fig. 3.12.

Zemel *et al.* [21, 38–40] compared the relative mobilities for a semiconductor film with a surface space charge region present both without and with an applied magnetic field. With no magnetic field [21],

$$\mu'/\mu_b = 1 - \exp(\alpha^2)\,\mathrm{erfc}(\alpha) \tag{3.47}$$

where $\alpha = L_c(2mkT)^{1/2}/q\tau V_s$. This expression is similar to that derived earlier by Petritz [41] for semiconductors and by Schrieffer [42] for metals.

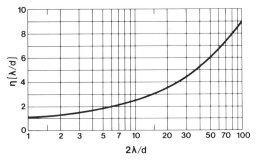

Fig. 3.12 The Hall coefficient correction factor as a function of the ratio of mean surface scattering length to film half-thickness (from Amith [37]).

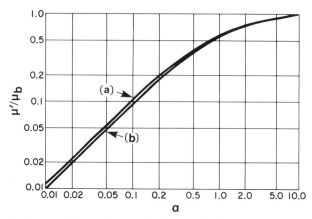

Fig. 3.13 The normalized film mobility (to the bulk value) as a function of the parameter α for (a) the effective conductivity mobility, and (b) the effective Hall mobility (from Zemel [21]).

With a magnetic field present, Zemel showed that [21]

$$(\mu'/\mu_b)_{R_H} = \left[1 - 2\alpha/\pi^{1/2} - (1 - 2\alpha^2)\exp(\alpha^2)\,\mathrm{erfc}(\alpha)\right]^{1/2} \quad (3.48)$$

where the subscript R_H indicates a Hall measurement. Equations (3.47) and (3.48) are compared as functions of α in Fig. 3.13, and the difference is quite small. For minimal band bending, V_s approaches zero and α becomes large. As expected, $\mu' = \mu_b = \mu'_{R_H}$ for this flat-band condition. As the surface potential well gets larger (i.e., V_s becomes large and surface band bending exists), α approaches zero. In this situation, $(\mu'/\mu_b)|_{\alpha \to 0} = 2\alpha/\pi^{1/2}$ and $(\mu'/\mu_b)_{R_H|\alpha \to 0} = \alpha$. Thus, even under this extreme condition, the mobilities differ by only about 12%.

Measurements of resistivity and the Hall coefficient are necessary for determining both the mobility and carrier concentration in thin films. However, caution must be exercised, since errors in measurements can arise from specimen contours; electrode size, geometry, position, and symmetry; and spatial and thickness inhomogeneities. Weider [43] reviewed these sources of error for galvanomagnetic measurements and presented a useful analysis that can be used either to avoid these errors or to correct for them.

5. Experimental Results

The semiconductor surface and its relationship to the electrical characteristics of the thin film have been the subject of many investigations. In general, such experiments are tedious and sometimes difficult to reproduce, especially if the films are polycrystalline. Although the thin films for such studies have been grown on a variety of substrates by all possible deposition techniques, epitaxial growth methods have been most successful for isolating the effects of surfaces, because of fewer defect-associated problems. With the improvements in deposition control, monitoring, and measurement for thin-film processing, it can now be expected that the semiconductor surface will be analyzed more accurately and rigorously. Among the more interesting and important are the recent advances in molecular beam epitaxy (MBE) [44–46]. This technique has the potential for making significant contributions to the knowledge of surface-related properties of films, since it produces very thin films, growing them a single layer at a time, with accurate and reproducible electrical, structural, and physical properties. It is *not* the purpose of this section to present an exhaustive compendium of surface-related data, but rather to describe results which verify or demonstrate the thin-film surface analyses and modeling discussed previously.

A. SURFACE SCATTERING

Evidence illustrating the effects of surface scattering upon the electrical properties of semiconductor and metal thin films is well represented in the literature [1, 36–43, 47–72]. The thickness dependence of the mobility (and resistivity) predicted by Eqs. (3.21), (3.22), and (3.29) has been experimentally demonstrated for Ge [40, 47–50], Si [36, 51–54], PbSe [38, 55], PbTe [39, 56], PbS [39, 57], CdS [29, 30, 58–63], CdSe [64], CdTe [65], GaAs [66, 67], and several other semiconductor thin films [68–72]. For the first example, Fig. 3.14 shows the dependence of the inverse film mobility upon thickness for CdS films deposited at two different substrate temperatures

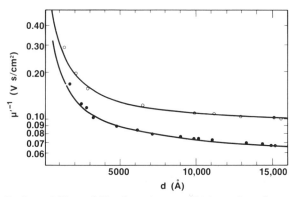

Fig. 3.14 Reciprocal film mobility dependence on thickness for substrate temperatures of 130°C (403°C) (open circles) and 180°C (453 K) (solid circles). The solid lines indicate the model of Eq. (3.29) with $\lambda = 1100$ Å for the CdS films. Deposition rate, 400 Å/min (from Kazmerski *et al.* [29]).

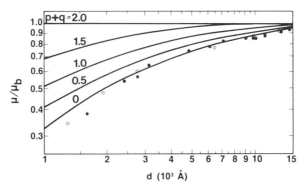

Fig. 3.15 Dependence of the film mobility on thickness for CdS thin films at substrate temperatures of 130°C (open circles) and 180°C (solid circles). The solid lines represent Eq. (3.30) for various values of the specular scattering parameters p and q, with $\lambda = 1100$ Å. The experimental data indicate that the scattering is almost entirely diffuse (from Kazmerski *et al.* [29]).

[29]. In each case the mobility approaches a constant value [μ_b in Eqs. (3.22) and (3.29)], while the mobility decreases significantly for films less than 1 μm in thickness. The solid lines represent the model of Eq. (3.29) with $\lambda = 1100$ Å. For these films, the scattering was found to be almost entirely diffuse by comparison to the relationship given by Eq. (3.30). Figure 3.15 shows that these data lie very near the $p + q = 0$ line [29].

The predicted dependence of the Hall coefficient (and carrier concentration) given by Eq. (3.45) has been verified for these CdS films. Figure 3.16 presents these data for a number of substrate temperatures [30]. The

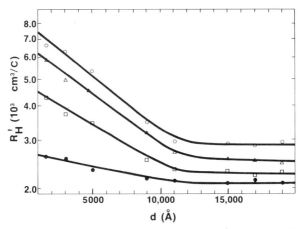

Fig. 3.16 Hall data dependence upon CdS film thickness for various substrate temperatures indicating variation of the mean surface scattering length with substrate temperature. Open circles, 220°C; open triangles, 180°C; open squares, 150°C; solid circles, 100°C (from Kazmerski [30]).

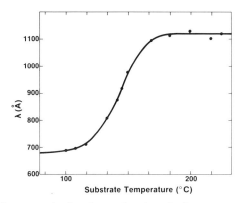

Fig. 3.17 Surface scattering length as a function of substrate temperature for CdS thin films (from Kazmerski [30]).

Hall coefficient for each case approaches a constant value for thicker films, with a relative increase for each substrate temperature observed for thinner films. By fitting these data in the vicinity of the knees of each of the curves in Fig. 3.16, the dependence of the surface scattering length upon substrate temperature has been evaluated (Fig. 3.17) [30]. The magnitude of λ varies from about 680 Å at low substrate temperature to 1200 Å at higher ones. The more or less constant λ at the extreme values is expected, since the carrier concentrations become constant in these ranges.

The effects of band bending upon the electrical properties of semiconductor films have been investigated by controlling the gaseous environment to which the film is exposed, thereby providing accumulation or depletion to some degree. Earlier studies exposed film surfaces to a variety of gases (e.g., O_2, H_2, H_2S) and correlated the changes in conductivity with the partial pressure of the gases [73–78]. These investigations are summarized in two good reviews [73, 74] and will not be covered in detail here. In some cases, the electrical properties were reversible with gas exposure and cycling, which led to the development of solid-state gas sensors [79].

B. FIELD-EFFECT EXPERIMENTS

Waxman *et al.* [32] demonstrated the effects of surface states on the mobility of CdS thin films by providing a solid interface at the CdS surface to control the band bending. In these experiments the band bending was varied by using a field-effect structure—similar to that of a field-effect transistor—which had a metal field-plate electrode deposited on the top insulator. A controlled bias V_p could be applied to the surface of the CdS, allowing observation of the relationship between change in band bending and the measured Hall mobility. Figure 3.18 shows the variation in Hall mobility with field-plate potential using a CaF_2 insulator on the CdS film

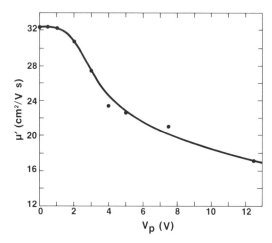

Fig. 3.18 Variation of Hall mobility with applied surface potential in field-effect structure with a 4000-Å CdS film and a 600-Å-thick CaF$_2$ surface insulator layer (from Waxman *et al.* [32]).

[78]. The CaF$_2$ tends to cause a *depletion region* at the surface of the CdS. The mobility decreases monotonically as a function of V_p, as predicted by Eq. (3.44), for increasing V_s (i.e., positive and increasing V_p). Figure 3.19 presents analogous data for a SiO–CdS structure for which the SiO tends to cause an *accumulation layer* in the CdS with no applied bias [78]. For $V_p < 6$ V, μ and the resistance are about constant, demonstrating the existence of surface states. The charge induced in these states is immobile, and the surface potential V_s is expected to be constant and independent of V_p; therefore, the mobility should be constant as predicted by Eq. (3.43). Above this 6-V region the ratio of free to trapped charge increases, and the surface mobility (and conductance) likewise increases. At very high V_p the mobility begins to decrease because of surface scattering.

Several other investigators used field-effect structures to study the mobility of semiconductor thin-film surfaces [80–83]. Van Heek [80] showed that even a relatively small number of surface states could have a strong influence on mobilities and carrier concentrations of CdSe thin films, consistent with the models discussed in the previous sections. Ipri used the field-effect technique to determine the electrical properties of Si films grown epitaxially on sapphire [81]. He varied the plate potential from negative through positive values, causing the Si film surface to change from depletion to flat-based to accumulation. The effect on mobility is shown in Fig. 3.20. A maximum is observed in this characteristic curve for light accumulation, indicating a higher mobility near the surface, and the mobility decreases at a higher V_p because of surface scattering. This result

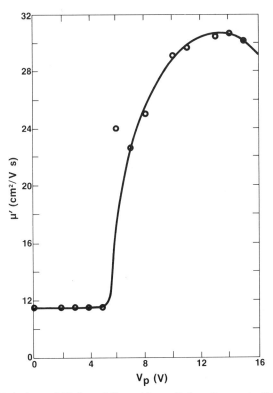

Fig. 3.19 Variatiom of Hall mobility with applied surface potential in field-effect structure with a 2000-Å CdS film and a 600-Å-thick SiO insulator layer (from Waxman *et al.* [32]).

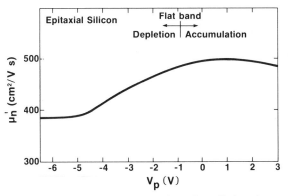

Fig. 3.20 Average film mobility as a function of applied surface potential for Si field-effect structure (from Ipri [81]).

is similar to that observed by Waxman (Fig. 3.19) for CdS films [32]. For the Si case, however, the mobility decreases rapidly as the depletion region widens, again indicating a mobility gradient through the film, consistent with the Anderson model (Fig. 3.10). For more negative values of V_p, the mobility becomes constant, since the inversion condition is reached and the depletion width becomes about constant. Ipri also provides a theoretical analysis with this field-effect technique to determine actual surface concentrations and the gradients of the mobilities and carrier concentrations through the z or thickness direction of the film.

3.3 TRANSPORT IN POLYCRYSTALLINE FILMS

Perfect epitaxial thin films are not the usual case encountered in thin-film technology. Even epitaxial films commonly contain point defects and dislocations that can affect carrier transport. Most often, the situation is even more complicated. For reasons of economics, large-scale deployment, or necessity, film growth is confined largely to a different material, noncrystalline substrates (e.g., on metal films for electrical contact), and the resulting semiconductor thin films are usually *polycrystalline*. Therefore, the conduction mechanism is dominated by the inherent *intercrystalline* boundaries rather than the *intracrystalline* characteristics.

This section focuses on the relationships between the defects found in polycrystalline semiconductor films and the resulting electrical characteristics. The intercrystalline boundaries, or *grain boundaries*, are emphasized because they are the most dominant, least controllable, and perhaps the most misunderstood problems for the thin-film investigator. Attempts at producing generalized models that explain the transport behavior due to grain boundaries have been unsuccessful. Although they provide insight into the general physics involved, the techniques must be understood as *modeling methods* which can predict behavior accurately only under a specific controlled set of experimental and material circumstances. It should be expected, for instance, that grain boundaries in compound semiconductor films are quite different for carrier transport than those in elemental semiconductors. Boundaries for larger crystallite films differ from those for small-grained ones. Also, the physical, structural, electrical, and optical characteristics of grain boundary regions are drastically altered by exposure to impurities, diffusion, and field effects. Thus, care must be taken in applying or interpreting any general analysis of grain boundary phenomena. Several models of polycrystalline semiconductor thin films are summarized here. Some expansions and explanations of the basic physical processes are included. Carrier transport associated with dislocations, stacking faults, mechanical stress, and defect types are discussed. Data representative of these effects are presented for elemental and compound semiconductors. Finally, the integration of surface scattering effects with defect-dominated phenomena is discussed.

1.   Initial Representation: Boundary Scattering

Figure 3.21 presents a conceptual cross section of a thin film having a cylindrical grain structure. This geometry has been observed in many thin-film systems and is effectively represented by the scanning electron micrograph of the cross section of a polycrystalline Si film shown in Fig. 3.22 [84, 85]. It should be added that, in some film growth cases, columnar growth can be interrupted and grain boundaries can occur along the z direction of the film.

A simple approach is to consider the grain boundaries represented in Fig. 3.21 as having the major effect of controlling carrier transport from grain to grain. Thus, the carriers collide at the boundaries and, in the steady state, have an effective mean free path λ_g and a mean relaxation

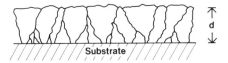

Fig. 3.21   Conceptual cross-sectional view of polycrystalline thin film indicating a degree of columnar growth.

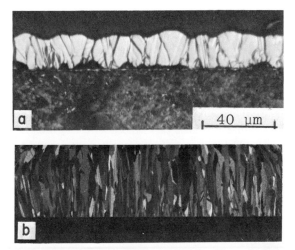

Fig. 3.22   Micrographs showing columnar growth in silicon. (a) Polycrystalline silicon film grown by chemical vapor deposition (From Chu *et al.* [84]). (b) Multicrystalline silicon produced by casting (from Helmreich [85]).

time τ_g. The mobility associated with this mechanism can be expressed

$$\mu_g = q\tau_g/m^* = q\lambda_g/m^*v \qquad (3.49)$$

where v is the mean thermal carrier velocity. Anderson [6] calculated this effective mobility for two cases. First, for a *degenerate semiconductor* (using Fermi–Dirac statistics),

$$\mu_g = (q\lambda_g/h)(3\pi^2 n)^{-1/3}(m_d^*/m_i^*)^{1/2} \qquad (3.50)$$

where n is the free carrier density, m_d^* the density of states effective mass, and m_i^* the inertial or conductivity mass of the carrier. Second, for a *nondegenerate* case (using Boltzmann statistics),

$$\mu_g = q\lambda_g(\tfrac{9}{8}\pi m_i^* kT)^{-1/2} \qquad (3.51)$$

The exact temperature sensitivities of Eqs. (3.50) and (3.51) will depend upon the semiconductor involved. If it has a low band gap, E_g is more affected by temperature changes (i.e., E_g becomes more significant), and m_i^* and m_d^* are especially temperature sensitive. Thus the mobility of a nondegenerate semiconductor is more likely to deviate from the $T^{-1/2}$ dependence of Eq. (3.51) for the low-band-gap case. On the other hand, the temperature sensitivity of the degenerate semiconductor mobility [Eq. (3.50)] depends upon the ratio of m_d^* to m_i^*. Since the temperature dependence of each of these masses is approximately the same, only small variations in μ_g with T are expected. For either case, the masses are relatively temperature-independent for materials with larger band gaps. Thus μ_g follows the $T^{-1/2}$ dependence for the nondegenerate semiconductor, and μ_g is relatively temperature-independent for the degenerate case. Measurements supporting this simple analysis are scant in the literature. In his work with PbTe, Egerton [86] reported the expected behavior for degenerate n-type films grown on mica. The mobility demonstrated was both reduced from the single-crystal value as predicted by Eq. (3.50) and approximately independent of temperature as expected for this low-band-gap semiconductor. This situation is somewhat special, since the boundary scattering model requires that the potential barriers at the grain boundaries be relatively small. The Egerton films fulfilled this condition, since the grain sizes were relatively small and the material permittivity was high, allowing little band bending to develop at the boundaries. This is, of course, similar to the surface scattering situation presented in Fig. 3.8 [34].

2. Grain Boundary Potential Barrier Models: Compound Semiconductors

Most analysis and modeling techniques correlating the transport properties with the polycrystallinity of thin films are based upon the consider-

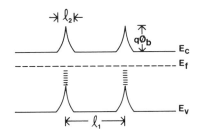

Fig. 3.23 Energy band representation of an *n*-type polycrystalline semiconductor thin film with grain size l_1 and grain boundary width l_2. Grain boundary barrier potential is $q\phi_b$.

ation that the grain boundaries have an inherent space charge region due to the interface. Band bending occurs, and potential barriers to the charge transport result. This situation is represented schematically for an *n*-type semiconductor in Fig. 3.23. The transport properties of both metals and semiconductors were scrutinized using this rather simple description, and the reduced mobility and conductivity of the materials as compared to single-crystal analogs were predicted. The major contributions to compound semiconductor thin films using this approach are now considered.

A. THE MODELS

Transport Mechanisms

One of the earliest models accounting for the conductivity in polycrystalline semiconductor films was that of Volger [87]. Volger based his model on that of an inhomogeneous conductor consisting of series-connected, separately homogeneous (electrically and structurally) domains of high conductivity and very low conductivity in which no space charge regions exist. The width of the low-conductivity domains (i.e., the grain boundary regions) is negligible with respect to the high-conductivity regions or grains. Thus, the situation simulates a polycrystalline semiconductor in which ohmic transport of the carriers dominates. Defining l_1 as the grain size and l_2 as the boundary width, Volger derived the expression for the Hall coefficient:

$$R_H = R_{H1} + c(l_2/l_1)^2 R_{H2} \qquad (3.52)$$

where c is an unspecified constant and R_{H1} and R_{H2} the Hall coefficients in the grain and boundary regions, respectively. He related this to the mobility, showing that

$$\mu_g = \mu_1\{[1 + (l_2/l_1)\exp(q\phi_b/kT)]^{-1} + (l_2/l_1)\} \qquad (3.53)$$

where μ_1 is the bulk grain mobility and ϕ_b a barrier potential relating to the concentrations in the grain and boundary domains.

This analysis was followed by perhaps the most cited theoretical analysis of transport mechanisms in polycrystalline thin films—that of Petritz [88]. Although his initial emphasis was directed toward the theory of photoconductivity in polycrystalline compound semiconductors, his straightforward modeling of the conductivity is applicable to a wider range of cases. The model differs slightly from Volger's and is based upon a polycrystalline semiconductor in which the thermionic emission of carriers is prevalent.

Petritz dealt with parameters as averages of many grains. Thus the analysis considered initially a single grain and a single boundary or barrier. The total resistivity of this case is written

$$\rho_g = \rho_1 + \rho_2 \tag{3.54}$$

where the subscripts signify grain or crystallite (1) and boundary (2) regions, respectively. It was assumed that, for the usual case, $\rho_2 \gg \rho_1$; then the current–voltage relationship for the barrier could be expressed (analogous to simple diode theory) as

$$j = Mn_1 \exp(-q\phi_b/kT)\left[\exp(-qV_B/kT) - 1\right] \tag{3.55}$$

where j is the current density, n_1 the mean majority carrier density in the grains, ϕ_b the potential height of the barriers, V_B the voltage drop across the barrier, and M a factor that is barrier-dependent but independent of ϕ_b. If the film has many barriers (i.e., small grain sizes), the voltage drop across any one is small as compared to kT/q, and Eq. (3.55) may be written

$$j = Mn \exp(-q\phi_b/kT)(qV_B/kT) \tag{3.56}$$

If V is the total voltage drop across the film and there are n_c crystallites or grains per unit length along the film of length L, then

$$j = \left[q\mu_0 n \exp(-q\phi_b/kT)\right]\mathcal{E} \tag{3.57}$$

where $\mu_0 = M/n_c kT$, $\mathcal{E}$ is the electric field, and the quantity in brackets is the conductivity. Petritz observed that the exponential term in Eq. (3.57) provided the essential characterization of the barrier. He assumed that, since $\rho_1 \ll \rho_2$, the carrier concentration was *not* reduced by the exponential factor, but rather that all carriers took part in the conduction process but with reduced mobility; that is,

$$\mu_g = \mu_0 \exp(-q\phi_b/kT) \tag{3.58}$$

Equation (3.58) can be generalized to include the case in which scattering within the grain can be significant [7]. To accomplish this, $\mu_0 = \mu_b(T)$, the bulk or single-crystal value. Therefore,

$$\mu_g = \mu_b \exp(-q\phi_b/kT) \tag{3.59}$$

It has been proposed that more correctly $\mu_b = C\mu_{cryst}$, where C is a constant and μ_{cryst} is the perfect crystalline value of mobility [29]. Thus μ_b is the bulk representation of the grain or crystallite mobility, which may include inherent defects or impurities.

Berger [62, 89] showed that the Hall coefficient and the carrier concentrations also exhibited exponential dependences, similar to that of Eq. (3.59). Berger showed that

$$R_H = R_0 \exp(-E_n/kT) \tag{3.60}$$

and the magnitude of the activation energy E_n depends upon the relative concentrations in the grain and the boundary region. Mankarious [90] and others [91–94] extended this work by observing that the conductivity term in Eq. (3.57) should be written more generally as

$$\sigma_g \sim \exp(-E_\sigma/kT) \tag{3.61}$$

where E_σ is the conductivity activation energy. Since $\sigma = ne\mu_n$ (for an n-type semiconductor) or $\sigma = pe\mu_p$ (for a p-type one), the carrier concentrations can be expressed by similar relationships, consistent with Berger's observation. Therefore,

$$n \sim \exp(-E_n/kT) \tag{3.62}$$

or

$$p \sim \exp(-E_p/kT) \tag{3.63}$$

where E_n and E_p are the carrier activation energies for n- and p-type films, respectively. The relationship among the conductivity, mobility, and carrier concentrations further predicts that

$$E_\sigma \simeq E_{n\,(or\,p)} + q\phi_b \tag{3.64}$$

Therefore, for the special case proposed by Petritz ($\rho_1 \ll \rho_2$), the mobility activation energy is identical to the conductivity activation energy, and R_H is constant.

Berger expanded the Petritz model further, using a more general yet precise representation of the polycrystalline film (Fig. 3.24). The grain size and resistivity are given by l_1 and ρ_1, and the boundary parameters by l_2 and ρ_2. Thus the effective resistivity is written

$$\rho_g = \rho_1 + (l_2/l_1)\rho_2 \tag{3.65}$$

Berger derived an expression for the effective mobility based upon this model:

$$\mu_g = \mu_1/\left[1 + c\exp(E/kT)\right] \tag{3.66}$$

where $c = (l_2/l_1)(n_{01}/n_{02})$ and n_{01} and n_{02} are the characteristic carrier concentrations at infinite temperature defined from $n_{1\,(or\,2)} = n_{01\,(or\,02)}$

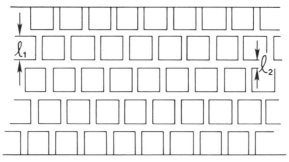

Fig. 3.24 Model of thin polycrystalline film proposed by Berger [89]. l_1 represents the grain size and l_2 the grain boundary width.

$\exp(-E_{1 \text{ (or 2)}}/kT)$ [see Eq. (3.62)], in the grain and grain boundary regions, respectively. μ_1 is the mobility in the grain. The second term in the denominator of Eq. (3.66) usually dominates, and the mobility term is often written in a form similar to that derived by Petritz:

$$\mu_g = \mu_1\left[(l_1/l_2)(n_{02}/n_{01})\exp(-\Delta E/kT)\right] \tag{3.67}$$

Comparison of Eqs. (3.58) and (3.67) leads to two interpretations of the activation energies and the magnitudes of μ_0 and μ_1. Kassing and Bax [95] demonstrated that, in the limiting (or comparative) cases for these two models,

$$\mu_0 \simeq \mu_1(l_1/l_2)(n_{02}/n_{01}) \tag{3.68}$$

and μ_1 is the mobility of the bulk single crystal. In the special case for which the grain size is large enough (i.e., greater than the mean free path of the carriers), scattering within the grain becomes important. Therefore, Eq. (3.68) reduces to $\mu_0 \simeq \mu_1 = \mu_b$, and Eq. (3.59) holds.

Barrier Heights

The barrier heights (or mobility activation energies) used in the previous models can be compared. For both the Volger and Petritz models [87, 88],

$$q\phi_b = kT\ln(n_1/n_2) \tag{3.69}$$

where n_1 and n_2 are the carrier densities in the grain and boundary regions, respectively. By comparison, Berger found that

$$\Delta E = E_2 - E_1 \tag{3.70}$$

where this difference is determined by the difference in the ratios:

$$\Delta E = \ln(N_A/N_D)|_{\text{grain boundary}} - (N_A/N_D)|_{\text{grain}} \tag{3.71}$$

and N_A and N_D are the acceptor and donor densities, respectively.

Since, in a more general sense than assumed by Petritz,

$$n_1 = n_{01} \exp(-E_1/kT)$$

and

$$n_2 = n_{02} \exp(-E_2/kT),$$

Eq. (3.69) becomes

$$q\phi = kT \ln(n_{01}/n_{02}) + (E_2 - E_1) \tag{3.72}$$

Substituting this into Eq. (3.58) shows that

$$\mu_g \sim (n_{01}/n_{02}) \exp\left[-(E_1 - E_2)/kT\right] \tag{3.73}$$

which is consistent with the Berger representation.

Other Analytical Approaches

Several other models for the conductivity in compound semiconductor films investigated produced results similar to those of Volger, Petritz, and

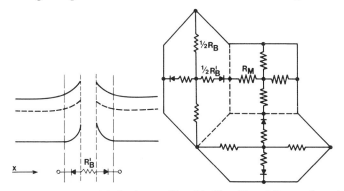

Fig. 3.25 Electrical model of polycrystalline thin film. R_B' and R_M are the resistances of the grain boundary and grain regions, respectively (from Kuznicki [96]).

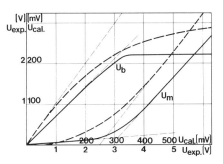

Fig. 3.26 Comparison of the experimental and calculated (using the model of Fig. 3.25) barrier voltage U_b and grain voltage U_M as a function of the total voltage V. Solid line indicates experimental data for CdSe, and the broken line indicates the calculated values (from Kuznicki [96]).

Berger [96–104]. Among them are several more complicated approaches based upon electrical component modeling of the polycrystalline film. An example is the methodology of Kuznicki, whose structural model and electrical equivalents are shown in Fig. 3.25. The model was used to verify numerically the static electrical characteristics of CdSe films. The qualitative resemblance of the analytical and experimental characteristics is shown in Fig. 3.26.

B. THE RESULTS

Experimental verifications of the models discussed in the previous section, especially that of the predicted reduced mobility and its exponential dependence upon inverse temperature, have appeared throughout the literature. Compound semiconductors reported to follow these models include: CdS [29, 32, 61, 103–115]; CdSe [64, 116, 117]; CdTe [118–121]; PbS [38, 122]; PbSe [38, 47, 52, 122–125]; PbTe [38, 55, 126–128]; InAs

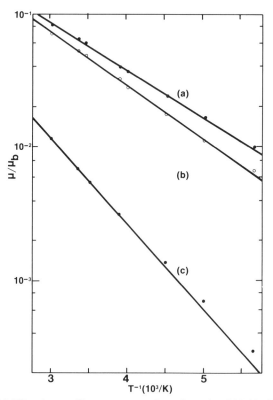

Fig. 3.27 Mobility–inverse film temperature dependence for CdS thin films at substrate temperatures. (a) 180°C (453 K; $q\phi = 0.071$ eV) and (b) 130°C (403 K; $q\phi = 0.076$ eV). (c) is identical to (a) but at a high deposition rate and $q\phi = 0.128$ eV (from Kazmerski et al. [29]).

[129]; InP [92, 130–133]; InSb [134–136]; Cu_xS [137–140]; SnO_x [141]; SnTe [39, 142]; GaAs [143–147]; GaP [148]; $CuInS_2$ [89, 149]; $CuInSe_2$ [90, 149–151]; $CuInTe_2$ [91].

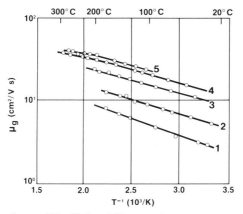

Fig. 3.28 Dependence of film Hall mobility upon inverse temperature for CdS thin films for various substrate temperatures: (1) 24°C (297 K); (2) 174°C (447 K); (3) 216°C (489 K); (4) 305°C (578 K); (5) 330°C (603 K) (from Berger [89]).

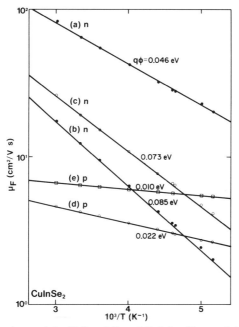

Fig. 3.29 Dependence of the Hall mobility of $CuInSe_2$ films on inverse temperature for various substrate temperatures: (a) 500 K, annealed 15 min in Ar; (b) As-deposited film, processed as in (a); (c) 520 K, As-deposited; (d) 520 K, annealed 15 min in Ar/H_2Se; (e) 520 K, annealed 80 min in Ar/H_2Se. Annealing temperature, 700 K (from Kazmerski *et al.* [91]).

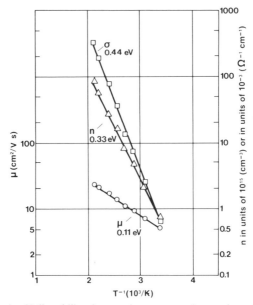

Fig. 3.30 Effective Hall mobility, free-carrier concentration, and conductivity of a CdS film as a function of reciprocal temperature (from Mankarious [90]).

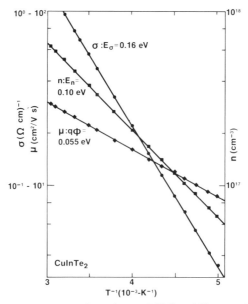

Fig. 3.31 Inverse temperature dependence of Hall mobility, conductivity, and carrier concentration for a p-type, annealed $CuInTe_2$ thin film. Substrate temperature is 473 K; thickness, 0.5 μm (from Kazmerski and Juang [93]).

Typical results are presented in Fig. 3.27 for CdS thin films deposited by resistive heating onto glass substrates. Values for $q\phi$ ranging from 0.05 to 0.50 eV have been reported for CdS films, and this activation energy can be seen from Fig. 3.28 for CdS and Fig. 3.29 for CuInSe$_2$ to be a function of the deposition conditions. The relationships among the mobility, conductivity, and carrier concentration activation energies are illustrated in Figs. 3.30 and 3.31. Figure 3.30 shows the data of Mankarious for CdS thin films, from which $E\sigma(0.44 \text{ eV}) = E_n(0.33 \text{ eV}) + q\phi_b(0.11 \text{ eV})$, as predicted by Eq. (3.64). This relationship has also been demonstrated for InP [92], CuInS$_2$ [89], CuInSe$_2$ [90], and CuInTe$_2$ (Fig. 3.31) [91].

Expected deviations from the exponential temperature dependence of electrical properties reported include nonuniform grain sizes [61, 152], concentration variations [153], and lattice and impurity scattering [39, 50, 88, 154]. For high temperatures Mankarious [88] demonstrated that the mobility of CdS films was dominated by both the impurity [see Eq. (3.5)] and lattice [see Eq. (3.4)] scattering (Fig. 3.32).

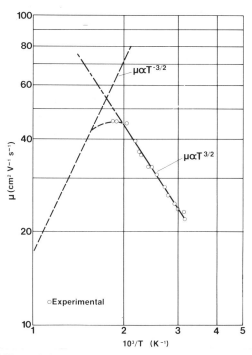

Fig. 3.32 Mobility variation as a function of temperature for a CdS thin film at elevated temperatures, indicating lattice scattering ($\sim T^{-3/2}$) and impurity scattering ($\sim T^{3/2}$) regimes (from Mankarious [90]).

3. Transport Properties: Elemental Semiconductors

The transport properties of elemental semiconductor films (Si and Ge) are considered separately from the previous section mainly because these semiconductors develop their intrinsic characteristics through the incorporation of impurity atoms rather than through compositional or stoichiometric effects. Two models have been used to interpret the electrical properties of elemental thin-film semiconductors:

(1) *The segregation model*, in which the impurity atoms segregate to the grain boundaries and are electrically inactive [155, 156]. The basic limitation of this model is that it fails to represent the temperature dependence of the resistivity. In addition, this model predicts a negative temperature coefficient of resistance which has not been experimentally demonstrable.

(2) *The grain boundary trapping model*, in which there is a large concentration of active trapping sites at the grain boundary which, in turn, captures free carriers [157–159]. As a result, the charge states at the grain boundaries become potential barriers as shown in Fig. 3.23. Similar to the situation in the Petritz and Berger models of the previous section, these barriers limit the transport of carriers between grains, and the mechanism is dominated primarily by thermionic emission.

Evolution of the potential at the grain boundary is an interesting phenomenon. In general, such grain boundary barriers are formed when the boundary region has a lower electrochemical potential for minority carriers than the grains, providing for the inflow of electrons or holes into the region. A space charge layer is thereby created that inhibits further flow of carriers. If barriers are formed in both n-type and p-type-majority carriers of a semiconductor, the Fermi level is located somewhere near the center of the band gap. Such is the case for silicon [156, 157]. However, the Fermi level is sometimes *not* in the midgap vicinity, and potential barriers can be created in only one majority carrier type of the semiconductor. Such is the case with germanium, in which grain boundary potential barriers form only in n-type material [160–162].

A. GRAIN BOUNDARY TRAPPING MODEL

Seto [163] developed the most comprehensive theoretical analysis of grain boundary trapping based upon the physical, charge, and energy band structures shown in Fig. 3.33. He made the following assumptions: (1) grains are identical, with cross section δ; (2) only one type of impurity atom is present (monovalent trapping) and is uniformly distributed with a concentration N/cm^3; (3) the grain boundary thickness is negligible and contains Q_t/cm^2 traps located at energy E_t with respect to E_i; (4) the traps

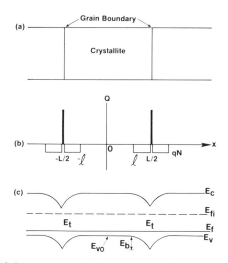

Fig. 3.33 Model for structure of polycrystalline film: (a) generalized structure; (b) charge distribution; (c) energy band structure (from Seto [163]).

are initially neutral and become charged by trapping a free carrier. Thus all the mobile carriers in a region $l/2 - \delta$ (see Fig. 3.33) from the grain boundary are trapped, resulting in a space charge region. Seto neglected the mobile carriers in this region.

The one-dimensional Poisson equation for this region, for a p-type film, is

$$d^2V/dx^2 = Nq/\epsilon, \qquad \delta < |x| < l/2 \qquad (3.74)$$

The solution of this equation yields, for $\delta < |x| < l/2$,

$$V(x) = (qN/2\epsilon)(x - \delta)^2 + V_{v0} \qquad (3.75)$$

where V_{v0} is the potential of the valence band edge (Fig. 3. 33).

Two possible conditions exist for the trap densities depending upon the doping concentrations: $Q_t > lN$ or $Q_t < lN$. For the former case, the grain is completely depleted of carriers and the traps are partially filled. Thus $\delta = 0$, and Eq. (3.75) yields the potential. Therefore, since $p(x) = N_v \exp[-(qV(x) - E_f)/kT]$, the average concentration, through integration over the region $-l/2 < x < l/2$, becomes

$$P_{av} = \left(\frac{N_v}{ql} \right)\left(\frac{2\pi\epsilon kT}{N} \right)^{1/2} \exp\left(\frac{E_b + E_f}{kT} \right) \text{erf}\left[\frac{ql}{2} \left(\frac{N}{2\epsilon kT} \right)^{1/2} \right] \qquad (3.76)$$

where n_i is the intrinsic concentration of the single crystal and

$$E_b = q\phi_b = q(ql^2N/8\epsilon) \qquad (3.77)$$

For $Q_t < lN$, only *part* of the crystallite is depleted of carriers and $\delta > 0$. In this case,

$$P_{av} = P_b\left[\left(1 - \frac{Q_t}{lN}\right) + \frac{1}{ql}\left(\frac{2\pi\epsilon kT}{N}\right)^{1/2}\right]\text{erf}\left[\frac{qQ_t}{2}(2\epsilon kTN)^{-1/2}\right] \quad (3.78)$$

where $p_b = N_v\exp[-(E_{v0} - E_f)/kT]$, as in similarly doped single-crystal silicon.

It should be noted that the net effect of doping concentration and trap density is to vary the barrier height $q\phi_b$. This behavior is shown for an arbitrary case in Fig. 3.34. The variation results from the dipole layer, which is created by impurities and filling of the traps. As the impurity concentration increases, so does the strength of the dipole region. Once all the traps are filled, however, both the width of the dipole layer and the magnitude of $q\phi_b$ decrease, while the total charge in the region remains constant.

The conductivity can now be determined using a thermionic emission current similar to that of Petritz [86]; that is,

$$j = qp_{av}(kT/2\pi m^*)^{1/2}\exp(-q\phi_b/kT)\left[\exp(qV_B/kT) - 1\right] \quad (3.79)$$

where V_B is the voltage applied across the grain boundary. For small V_B ($\ll kT/q$), the conductivity is calculated from Eq. (3.79):

$$\sigma_g = q^2lP_{av}(2\pi m^*kT)^{-1/2}\exp(-q\phi_b/kT) \quad (3.80)$$

Thus two solutions are obtainable, corresponding to the doping density regimes. Substituting Eqs. (3.76) and (3.78) into Eq. (3.80) yields

$$\sigma_g \sim \exp\left[-(E_g/2 - E_f)/kT\right], \quad lN < Q_t \quad (3.81)$$

and

$$\sigma_g \sim T^{-1/2}\exp(-E_b/kT), \quad lN > Q_t \quad (3.82)$$

In either case, the effective mobility from Eq. (3.80) is

$$\mu_g = ql(2\pi m^*kT)^{-1/2}\exp(-q\phi_b/kT) \quad (3.83)$$

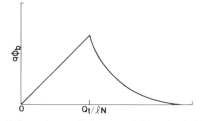

Fig. 3.34 Functional dependence of the potential barrier height on doping concentration; $N_t = 3.8 \times 10^{12}$ cm^{-2}; $L = 3 \times 10^{-6}$ cm (from Seto [163]).

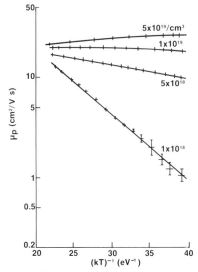

Fig. 3.35 Film Hall mobility as a function of inverse temperature for various doping concentrations in polycrystalline Si (from Seto [163]).

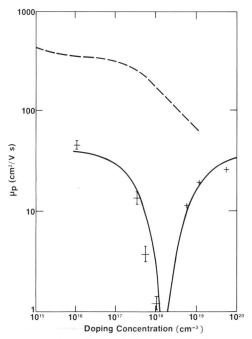

Fig. 3.36 Room-temperature Hall mobility versus doping concentration for p-type polycrystalline Si. The experimemtal results (+) are compared with the theoretical solid curve. The broken line is for single crystal Si (from Seto [163]).

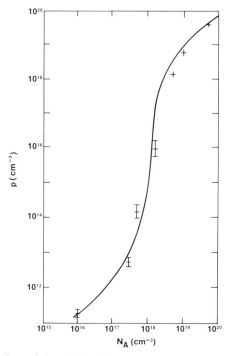

Fig. 3.37 Comparison of the calculated average carrier concentration as a function of doping concentration from room-temperature Hall data for polycrystalline Si. The experimental results (+) are compared with the theoretical solid curve (from Seto [163]).

In the case of polycrystalline Si, Seto showed experimentally the validity of Eqs. (3.81) and (3.82) by plotting the logarithm of the resistivity versus $(kT)^{-1}$ [158]. These data gave the predicted straight-line dependence with slope $E_g/2 - E_f$ for $lN < Q_t$ and E_b for $lN > Q_t$. Figure 3.35 presents the data for the mobility of Seto's polycrystalline Si films as a function of inverse temperature. As expected, the dependence shows a straight line with a negative slope $(= q\phi_b)$ and a deviation when $q\phi_b < kT$. Seto's model also accounts for the Hall mobility, carrier concentration, and resistivity dependencies upon doping concentration for polycrystalline films (Figs. 3.36–3.38). In each, the solid line indicates the model with a very good correspondence observable.

B. MODEL LIMITATIONS AND REFINEMENTS

Seto [159, 163] and others [164–166] noted several basic limitations of the grain boundary trapping model, including the following.

(1) *Grain resistivity.* In Seto's model, the bulk resistivity of the grain was assumed to be insignificant with respect to the resistivity of the

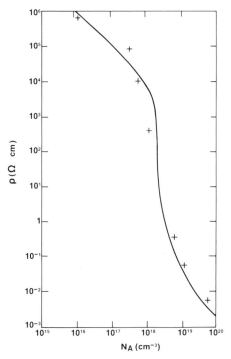

Fig. 3.38 Room-temperature resistivity as a function of doping concentration for polycrystalline Si. The experimental results (+) are compared with the solid line representing the theoretical analysis (from Seto [163]).

boundary region. If the grain size is large and the doping relatively high, the grain resistivity must be considered. Kamins demonstrated this effect with polycrystalline Si doped above $7 \times 10^{18}/\text{cm}^3$ [157, 167, 168].

(2) *Discrete versus distributed energy states.* Rather than the discrete or fixed energy assumed in the model, it is possible that the trapping states at the grain boundaries are distributed over some energy range, as reported for the surface states of the free Si surface [169] or the Si–SiO$_x$ interface [170]. This affects directly the activation energy, mobility, and carrier concentration, especially if $N \lesssim Q_t/l$ (see Fig. 3.34).

(3) *Depletion region free-carrier density.* In some cases (e.g., large-grain polycrystalline Si), the carrier concentration in the space charge layer can become appreciable, leading to inaccurate values of the barrier heights calculated by the Seto model. Since μ_g depends exponentially upon $q\phi_b$, significant variances can result. The carrier concentration is more sensitive to the shape of the barrier in the depletion region and is less affected by changes in the magnitude of the barrier height.

(4) *The energy level of interface (trapping) states.* The Seto model predicts (depending upon impurity type) that the trapping states are

located in either the upper or lower half of the band gap—excluding the midgap position [164, 166].

(5) *The trap population.* A major assumption of the Seto model, that available grain boundary traps in the band gap are always filled, is not universally true.

Several modifications of the Seto model have been proposed by Baccarani *et al.* [164]. Directed toward clarifying the model for the intermediate range of impurity concentrations, their work considers two trapping cases: (1) monovalent trapping states, and (2) continuous distribution of trapping states within the band gap.

Monovalent Trapping States at the Grain Boundary

For the case of monovalent trapping states at the grain boundary, Baccarani *et al.* assume the existence of N_t acceptor states (using an *n*-type semiconductor analogous to the Seto *p*-type situation) with energy E_t with respect to E_i at the interface. For a given set of l, N_t, and E_t there is an impurity concentration N_D^* such that, if N_D *is less than* N_D^*, the grains are entirely depleted. Two possible impurity conditions exist:

(i) For $N_D < N_D^*$, the energy barrier is given by [164]

$$E_b = q^2 l^2 N_D / 8\epsilon \qquad (3.84)$$

and the conductivity becomes

$$\sigma_g = \left[q^2 l^2 N_c N_D v / 2kT(N_t - lN_D) \right] \exp(-E_a/kT) \qquad (3.85)$$

where $v = (kT/2\pi m^*)^{1/2}$ and the activation energy is given by

$$E_a = E_g/2 - E_t \qquad (3.86)$$

(ii) For $N_D > N_D^*$, the grains are partially depleted and the barrier height is [164]

$$E_b = E_f - E_i + kT \ln \left[2(qN_t/(8\epsilon N_D E_b)^{1/2} - 1) \right] \qquad (3.87)$$

which must be solved iteratively. Two solutions for the conductivity can be calculated corresponding to two energy regions:

$$\sigma_g = (q^2 l N_c^2 v n_0 / kT) \exp(-E_a/kT),$$
$$E_f - E_t - E_b \gg kT \quad (3.88)$$
$$E_a = E_b,$$

and

$$\sigma_g = qN_c^2 v (2\epsilon N_D^{-1} E_b)^{1/2} (kTN_t)^{-1} \exp(-E_a/kT),$$
$$E_t + E_b - E_f \gg kT \quad (3.89)$$
$$E_a = E_g/2 - E_t,$$

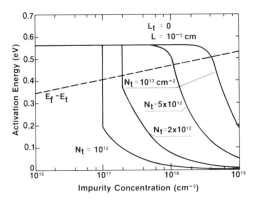

Fig. 3.39 Activation energy as a function of impurity concentration for polycrystalline Si (grain size, 10^{-5} cm). Reference is made to monovalent trapping states at midgap (from Baccarani *et al.* [164]).

where n_0 is the concentration in the neutral region neglected by the Seto treatment. The latter energy region [Eq. (3.89)] also had not been covered by Seto's approach and solution.

Figure 3.39 represents the calculated activation energy as a function of N_D for several trap densities. The grain sizes are assumed to be 0.1 μm, and the dashed line indicates the boundary between the two energy regions and their respective solutions [Eqs. (3.88) and (3.89)]. For lower values of N_t, an abrupt transition in E_a occurs (between $E_g/2$ and E_b) at the onset of complete depletion. However, for larger N_t, E_a becomes more continuous. For either complete or incomplete depletion, E_a approaches $E_g/2 - E_f$ for the condition given by Eq. (3.87).

Continuous Distribution of Trapping States

For the comparative case of a continuous energy distribution of interface states, N_i (expressed in units of inverse square centimeters per electron volt), acceptors are assumed to be uniformly distributed in the upper half of the band gap and donors in the lower half. Baccarani *et al.* solved the charge neutrality equation for this situation to determine the effective space charge region width and the position of the Fermi level in terms of the barrier potential, doping, grain size, and interface state density. For *complete depletion* of the grains ($N_D < N_D^*$) [164],

$$\sigma_g = (q^2 l N_c v / kT) \exp(- E_a / kT) \qquad (3.90)$$

with

$$E_a = E_g/2 - l N_D / N_i \qquad (3.91)$$

Corresponding to *incomplete depletion* ($N_D > N_D^*$ and the space charge

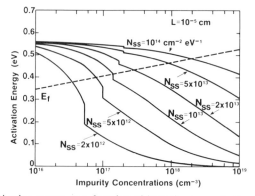

Fig. 3.40 Activation energy as a function of impurity concentration for polycrystalline Si (grain size, 10^{-5} cm). Reference is made to a distribution of interface states (from Baccarani *et al.* [164]

region width $< l/2$) [164],

$$\sigma_g = (q^2 l N_c v / kT)(N_c/N_D)^{-\beta} \exp(-E_a/kT) \qquad (3.92)$$

with

$$E_a = E_g/2 - \left[(4\epsilon N_D)/(q^2 N_i)^2 \right](\beta - 1) \qquad (3.93)$$

where $\beta = (1 + q^2 N_i^2 E_g / 4\epsilon N_D)^{-1/2}$.

Figure 3.40 shows the calculated activation energy versus N_D for the distributed interface state case with the same basic parameters used to determine Fig. 3.39, the comparable discrete situation. For smaller impurity densities E_a approaches $E_g/2$, and for higher impurity concentrations it is proportional to N_D^{-1}. Both these limits correspond to monovalent trapping states. However, the transition between these two limits is more

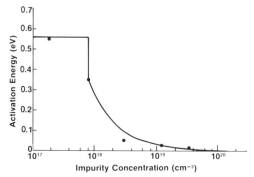

Fig. 3.41 Theoretical and experimental activation energy as a function of impurity concentration in polycrystalline Si (from Baccarani *et al.* [164]).

extended in the present case, with the abrupt transition between complete and partial depletion always occurring, even at high N_i.

Baccarani *et al.* performed Hall studies on sputter-deposited polycrystalline silicon films implanted with phosphorus. Their results indicate some impurity segregation at the grain boundaries and the presence of an interface state peak located at *midgap*. Figure 3.41 presents their activation energy data taken as a function of phosphorus concentration. The solid line indicates the monovelent trapping state model fitted to the experimental data by adjusting N_t to $3.9 \times 10^{12}/\text{cm}^2$. The model based upon the continuous distribution of interface states failed to fit the experimental data.

c. Grain Boundary Potential: Concentration Problems

Three major problems associated with the quality and nature of polycrystalline semiconductors—especially Si—have impeded the exact interpretation of the results of transport studies. The first is the uncertainty in the actual total doping of films. This is especially true of the chemical vapor deposition process, in which uniform accurate control is difficult, although Monkowski *et al.* [171] recently reported that dopant uniformity in their films matches single-crystal uniformity. However, even in the case of ion-implanted films, such as those used by Baccarani *et al.* [164], the postimplantation high-temperature anneals necessary to minimize high-energy damage can cause impurity diffusion, especially to grain boundary regions. The second problem is the uncertainty in impurity concentration, which varies with film thickness [157, 172]. The third problem concerns contamination of the grain boundary by gases inherent to the film growth procedure. Although film-processing techniques can minimize this risk, the large grain boundary surface-to-volume ratios in polycrystalline thin films magnify the problem.

In order to avoid such complications, Seager and Castner [165] characterized the electrical transport properties of neutron transmutation-doped polycrystalline Si [173]. Their samples had larger grain sizes, with a minimum of 25 μm diameter. The electrical measurement techniques included four-probe and traveling potential probe measurements carried out as close to zero bias as possible. For very uniform controlled samples, Seager and Castner observed that the resistivity was linearly dependent upon inverse temperature—as previously observed [157–159, 163]—below a doping level of approximately $2 \times 10^{15}/\text{cm}^3$ of P. The activation energy was nearly the midgap value, 0.55 eV. However, above $2 \times 10^5/\text{cm}^3$ doping, deviations from linear dependence were observed (Fig. 3.42). Potential probe measurements made on these more highly doped polycrystalline Si samples indicate that a large range of grain boundary

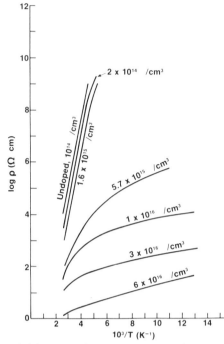

Fig. 3.42 Log resistivity versus inverse temperature for neutron transmutation-doped polycrystalline Si (from Seager and Castner [165]).

impedances exist (i.e., a variety of grain boundary barrier heights are present).

Seager and Castner examined the mechanism for current flow in the presence of grain boundary potential barriers. The grain boundary was represented analytically under three different conditions for the interface or defect state density N_i:

(1) *Energy-independent* N_i:

$$q\phi_b = \Delta E_f\left[1 + \alpha/2\,\Delta E_f - \left(\alpha/\Delta E_f + \alpha^2/4\,\Delta E_f^2\right)^{1/2}\right] \qquad (3.94)$$

where $\Delta E_f = E_{fg} - E_{fb}$, the Fermi levels in the grain and boundary regions, respectively, measured from the valence band edge, and $\alpha = 8\epsilon N_D/q^2 N_t$, with N_D the doping level and N_t the two-dimensional density of defect states.

(2) *Monoenergetic* N_i: In this case, a closed form solution is not obtained, but

$$\left[8\epsilon N_D q\phi_b/\left(q^2 N_t^2\right)\right]^{1/2} = \left(\alpha q\phi_b\right)^{1/2} \qquad (3.95)$$

which can be solved using iterative techniques.

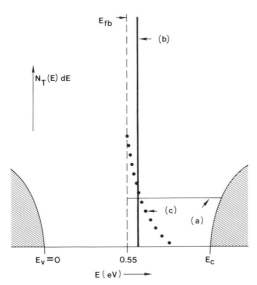

Fig. 3.43 The density of states and Fermi level in the barrier region for polycrystalline Si: (a) $N_t = 3.3 \times 10^{12}/\mathrm{cm}^2$ eV, (b) $N_t = 5.4 \times 10^{11}/\mathrm{cm}^2$ eV at 0.626 eV above E_v and zero elsewhere, and (c) $N_t = (1.89 \times 10^{15}/\mathrm{cm}^2$ eV$)\exp(-E/0.099$ eV$)$ (from Seager and Castner [165]).

(3) *Exponentially dependent* N_i: A transcendental equation also results:

$$\left(8\epsilon N_D/q^2\right)^{1/2}(q\phi_b)^{1/2} = E_0\left[N_{TO}\exp(-E_{fb}/E_0)\right]$$
$$\times\left\{1 - \exp\left[(q\phi_b - \Delta E_f)/E_0\right]\right\} \quad (3.96)$$

where $N_t = N_{TO}\exp(-E_{fb}/E_0)$ and N_0 and E_0 are adjustable parameters.

These three cases for N_i are summarized in the band diagram in Fig. 3.43 with the parameters used in Eqs. (3.94)–(3.96) to fit the zero-bias data specified. Figure 3.44 presents the dependence of $q\phi_b$ upon N_D for the three interface density cases, using the fitting parameters determined by Seager and Castner [165]. The exponential density-of-states model fits the data somewhat better than the single-trap model, and the energy-independent N_i model has the poorest agreement. These comparisons indicate that the largest grain boundary state densities consist of $6 \times 10^{11}/\mathrm{cm}^2$ available electron states located within 0.2 eV of the midgap position. This is lower than the corresponding value of $3 \times 10^{12}/\mathrm{cm}^2$ in the Seto and Baccarani polycrystalline Si [163, 164], which is ascribed to the extrinsic quality (probably due to contamination) of the grain boundary states.

Seager and Castner correlated their results with the effects of doping

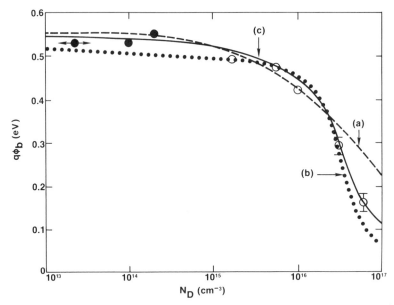

Fig. 3.44 The largest grain boundary barrier heights ($T = 0$) as a function of doping density for neutron transmutation-doped polycrystalline Si. Open circles are data from small-area potential profile experiments, and solid circles are four-probe resistivity data. Dashed curve (a), uniform density of states; dotted curve (b), single trap level; solid curve (c), exponential density of states (from Seager amd Castner [165]).

concentrations on the electrical transport in polycrystalline Si. The regions of interest, based upon Fig. 3.42, include:

(1) $N_D < 10^{14}/cm^3$: The majority of the barriers in polycrystalline Si are characterized by $q\phi_b = 0.55$ eV, and the resistivity is dominated by this activation energy.

(2) $10^{14} < N_D < 2 \times 10^{15}/cm^3$: A substantial number of barrier heights are less than 0.55 eV, but the largest $q\phi_b$ still dominate the resistivity–temperature dependence.

(3) $N_D > 2 \times 10^{15}/cm^3$: A range in barrier heights exists, and the magnitude of the resistivity activation energy depends upon the physical features of the sample and the analysis technique. A major problem concerning the shape of the grain boundary density-of-states function still exists. Seager and Castner propose that a careful (but difficult) measurement of the I–V characteristics of individual grain boundaries would provide this information.

The modeling work of Seager and Castner [165] and the application studies of Seager, Ginley and Pike [174–178] have provided important contributions to the understanding of both the electrical behavior of grain boundaries and their ultimate control in polycrystalline devices. Their data demonstrate the effectiveness of the double-depletion layer–thermal emission model for the formation of and electrical transport across silicon grain boundaries. A major conclusion of this work is that the barrier height is temperature dependent, a fact not usually recognized in grain boundary and polycrystalline semiconductor electrical formulations. This effective activation energy is more accurately expressed at zero bias [165]:

$$E_a = q[\phi_b - T(\partial\phi_b/\partial T)]$$ (3.97)

The correctness of this temperature dependence is illustrated in Fig. 3.45 for polycrystalline Si [177, 178]. The barrier height data follow a linear dependence upon temperature, in this case with a slope of 1.23 meV/K. The quantity commonly deduced from experimental measurements is the conductivity activation energy, given by Eq. (3.97). As long as the doping is in a suitable range, the magnitude of E_a is essentially independent of temperature as shown in Fig. 3.45. However, if this suitable range is exceeded, the energy dependence of the grain boundary interface state density affects ϕ_b, and the second term of Eq. (3.97) is no longer nonnegligible. Seager and Ginley have demonstrated the importance of such

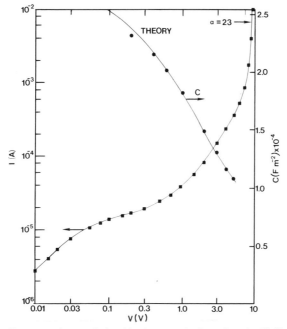

Fig. 3.45 Current–voltage relationship for a grain boundary in Si illustrating ohmic, saturation, and varistor regimes. Circles, high-frequency (10^6 Hz) capacitance per area ($T = 302$ K); squares, dc current ($T = 310$ K). $N_d = 10^{22}$ m^{-3}; barrier area, 1.3×10^{-6} m^2 (from Seager and Ginley [177]).

changes in interface state densities by hydrogenation of the grain boundary regions, which has directly led to consideration of this process for selective passivation of these problem areas [174, 178].

Recently, Seager, Pike, and Ginley have further added to the understanding of grain boundary transport mechanisms by evaluating and measuring the $I-V$ relationships across individual grain boundaries [174, 178]. Their data on silicon indicate that the boundary response can be divided into three voltage-dependent regions. Defining the parameter

$$\alpha_{gb} = d(\ln J)/d(\ln V) \tag{3.98}$$

the regions can be defined as

Ohmic: $\alpha_{gb} = 1,$ $V < kT/q$
Saturation: $\alpha_{gb} < 1,$ $kT/q < V < V_s$
Varistor: $\alpha_{gb} > 1,$ $V > V_s$

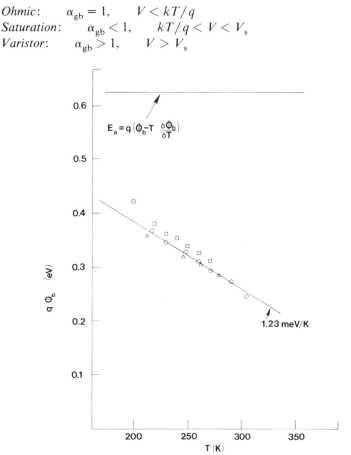

Fig. 3.46 Grain boundary barrier height as a function of temperature for polycrystalline Si. The dependence of E_a (Eq. 3.97) is also shown. Solid circles, four-probe data; open circles, potential profile data. (a) Uniform density of states; (b) single-trap level; (c) exponential density of states $N_d = 1.43 \times 10^{15}$ cm^{-3}. ○, 18 Å; Δ, 13Å; □, 15 Å (from Seager and Ginley [177]).

where V_s is defined as the voltage at which $a = 1$. It is noted that the varistor region is due to the lowering of ϕ_b. Supporting $I-V$ data are presented in Fig. 3.46 [177, 178]. The ohmic region, essentially a straight-line dependence of I upon V, is observed for $V < 0.3$ V. The saturation region is evident until the current rises steeply with increasing voltage, indicating the onset of the varistor region.

The work of Seager, Castner, and Ginley supports the importance of considering the chemistry of grain boundaries for their complete electrical characterization. In fact, this local chemistry can be the controlling factor, since even small differences in the chemical and compositional makeup of these regions can vastly alter the grain boundary interface state densities. Recent surface analysis results enforce this proposition, since chemical differences have been discovered not only from grain to grain in a given, as-produced polycrystalline sample, but also the localization of impurities along a single grain boundary has been reported [179–181]. These results could greatly complicate the generalization of any modeling technique.

4. Grain Boundary Measurements

Although some research has focused directly on the electrical and associated properties of individual grain boundaries [161, 182–189], the literature—including a detailed overview by Matare [182]—has been scant in synthesizing such data with either the modeling results or the broader application aspects of polycrystalline films. However, it is becoming more evident that such research and correlations are now necessary, especially with the potential markets provided by the large-scale deployment of thin-film solar cells and other polycrystalline semiconductor devices [190, 191].

Cohen *et al.* reported on the carrier transport at twin and low-angle boundaries in MBE-produced GaAs [192]. This work is representative of the synergistic approach mentioned above. They reported the measurement, by scanning Auger spectroscopy, of the resistivity of single-grain and twin boundaries in controlled GaAs samples based upon spatially resolved potential techniques.

In their experiment, two ohmic contacts were deposited such that a single-grain boundary was located perpendicular to them (Fig. 3.47). Auger electron spectroscopy is ordinarily used as a surface chemical analysis technique, since the escape depths of the analyzed Auger electrons usually are less than 10 Å [193]. However, Cohen *et al.* utilized this surface constraint along with the fact that the emitted Auger electron has a characteristic energy relative to the potential of the host atom with the analyzer referenced to ground. If the sample is biased, however, the entire Auger spectrum is shifted by an energy proportional to the known bias voltage [192]. Cohen *et al.* applied a voltage between the two ohmic

Fig. 3.47 Electron micrograph of a 20-μm-long GaAs sample containing a single grain boundary (from Cohen *et al.* [192]).

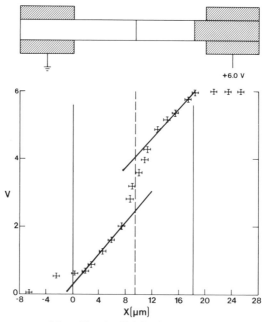

Fig. 3.48 Auger potential profile of a lightly doped GaAs sample containing a single grain boundary (from Cohen *et al.* [192]).

contacts and used the energy shift of the carbon Auger peak (272 eV) as a spatially resolved contactless voltmeter [192]. The carbon line was chosen for its sharpness and the fact that the element occurs on any surface exposed to air.

Figure 3.48 shows the result for a lightly doped sample and a 6-V bias with a resulting 6×10^{-4}-Å current. From the grounded contact, the voltage increases linearly across the sample as far as the vicinity of the grain boundary. In this region an observable change in slope, representative of the grain boundary resistance, is seen. Beyond this region, the slope continues to increase linearly with its original slope.

The increase in voltage at the grain boundary in Fig. 3.48 corresponds to a region more than 3 μm in width, which represents the depletion width of a diode fabricated from such lightly doped material. Cohen *et al.* note that the Fermi level of the GaAs is probably pinned at the grain boundary by interface states. Band bending on both sides of the interface would

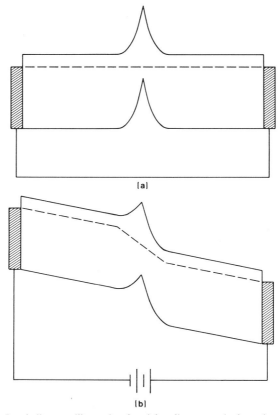

Fig. 3.49 Band diagram illustrating band bending at grain boundary in GaAs (a) in equilibrium, and (b) with applied voltage (from Cohen *et al.* [192]).

result, forming a double-depletion region (shown in Fig. 3.49 When a potential is applied to the sample, the major portion appears across the depletion region at the grain boundary as long as the grain resistivity is moderate. The resistivity of the grain boundary shown in Fig. 3.48 was calculated to be $3.3 \times 10^{-4}\Omega$ cm^2, which is consistent with that reported by Dapkus *et al.* [194].

5. Generalized Hall Parameters for Semiconductor Thin Films

Hall-effect measurements are the most common technique used in determining thin-film mobility and carrier concentration. Several interpretive problems that can arise from such measurements have been discussed by Seager and Castner [165]. Recently, a phenomenological theory of the Hall effect in polycrystalline semiconductors, providing for a general and reasonable interpretation of results, was reported by Jerhot and Snejdar [195]. Their approach models the film as depicted in Fig. 3.50, which also shows that the detailed electrical equivalent circuit of the film is compatible with previous analogies [94–99]. Jerhot and Snejdar provided a complex analysis of this model, with the resultant Hall coefficient given by

$$R_{Hg} = \left[\sigma'(d_G + d_B) \right]^{-1} \Bigg(R_{HBb}\sigma_{Bb}d_B$$

$$\times \left\{ \frac{d_G - d_B}{2L_{Db}d_G} (R_N\sigma_{Bb}d_B)^{-1}[J_2(V_{Db}, V_{0b})]^{-1} + 1 \right\} + R_{B2}D \Bigg) \quad (3.99)$$

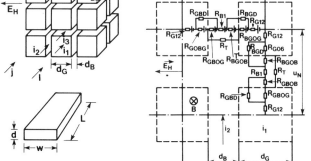

Fig. 3.50 Assumed structure of polycrystalline semiconductor for generalized Hall-effect analysis and equivalent electrical circuits. Subscripts G, B, and b refer to grain, grain boundary, and bulk, respectively (from Jerhot and Snejdar [196]).

and the Hall mobility by

$$\mu_g = (d_G + d_B)^{-1} \left\{ R_{HBb}\sigma_{Bb}d_B \right.$$

$$\times \left[\frac{d_G + d_B}{2L_{DB}d_G} (R_n\sigma_{Bb})^{-1} \frac{1}{J_2(V_{DB}, V_{0B})} + 1 \right] + \left. R_{B2}D \right\} \quad (3.100)$$

where the various parameters are defined in Fig. 3.50 and Table 3.1.

This generalized approach can be used to generate the analytical results presented previously [85–87, 158, 196]. Jerhot and Snejdar havs shown that the proper expressions for μ_g and R_{Hg} are obtained from Eqs. (3.99) and (3.100) for the limiting cases derived by Volger, Petritz, Berger, and Seto [85–87, 158]. In line with previous discussions of the Hall effect [43, 196], they conclude the following from their modeling analysis.

(1)   In polycrystalline semiconductors, the Hall mobility measured should never be equated with the conductivity mobility [86] even if the Hall coefficient is known. Thus, the physical parameters associated with the charge transport cannot be calculated directly from μ_H measured under the condition of zero current through the Hall contacts.

(2)   In general, a smaller value of Hall mobility is measured in polycrystalline semiconductors as a result of the smaller number of carriers that transfer charge between current contacts as compared with the potential barrier-free case. Their predicted barrier height effects on mobility and carrier concentration are shown in Fig. 3.51.

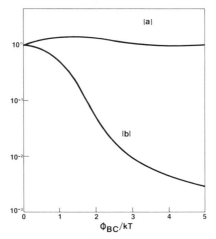

Fig. 3.51   Effect of barrier height on (a) Hall coefficient (R_{Hg}/R_{HBb}) and (b) Hall mobility (μ_{Hg}/μ_{HBb}) in polycrystalline semiconductor, with $n_b = 10^{16}/cm^3$, $l_1 (= d_g) = 10^{-5}$ cm, $l_2 (= d_b) = 10^{-7}$ cm, and $\mu_b = 10$ cm^2/V sec (from Jerhot and Snejdar [196]).

TABLE I

Summary of Generalized Hall-Effect Parameters for a Polycrystalline Semiconductor [a,b]

$d_G = l_1$ = grain size	$d_B = l_2$ = grain boundary width
σ_p = conductivity of polycrystalline semiconductor	σ_{Bb} = bulk conductivity of grain boundary region
R_N = resistance of one current path formed by interconnected grains and grain boundary regions (see Fig. 3.48)	L_{DB} = effective Debye length in grain boundary region

$$J_i(V_1, V_2) = -(q/kT)\int_{V_1}^{V_2}[\exp(qV/kT)/F_1(V, V_{0B})]\,dV, \quad \text{where} \quad i = 1 \text{ or } 2$$

$$F_1(V, V_{0B}) = \pm\left\{2[\exp(qV/kT) - \exp(qV_{0B}/kT) + (q/kT)(V_{0B} - V)]\right\}^{1/2},$$

where + corresponds to $V < 0$ and − corresponds to $V > 0$

$$D = D_u/D_s$$

$$D_u = \det(A_{1n}, R_{m,n}), \quad \text{where} \quad m = 2, 3, 4 \text{ and } n = 1, 2, 3, 4$$

$$D_s = \det(R_{m,n}), \quad \text{where} \quad m = 1, 2, 3, 4 \text{ and } n = 1, 2, 3, 4$$

$R_{m,n}$ is a symmetrical square matrix with

$$R_{1,1} = R_G + 2(R_{GBOG} + R_{GBOB}) + R_{B1} + R_{B2};$$

$$R_{1,2} = R_{2,1} = R_{1,3} = R_{3,1} = -(R_{GBOG} = R_{GBOB})$$

$$R_{1,4} = R_{4,1} = -(R_{B1} + 2R_{GBOB}); \quad R_{2,2} = R_{2,3} = R_{GBOG} + R_{GBOB} + R_{GBD}$$

$$R_{2,3} = R_{3,2} = 0, R_{2,4} = R_{4,2} = R_{3,4} = R_{4,3} = R_{GBOB}; \quad R_{4,4} = R_T + R_{B1} + 2R_{GBOB}$$

$A_{1,n}$ elements are

$$A_{1,1} = R_{HGb}[2L_{DG}(d_G + d_B)/R_N d_G^2]J_{1G}(V_{DG}, V_{0G}) + R_{HBb}\sigma_{Bb}d_B$$

$$\times \left\{(2L_{DB}/d_B)^2 J_{1B}(V_{DB}, V_{0B})J_2(V_{DB}, V_{0B})\right.$$

$$\left. - (d_G + d_B)/[2L_{DB}R_N\sigma_{Bb}d_G J_2(V_{DB}, V_{0B})] - 1\right\}$$

$$A_{1,2} = A_{1,3} = -\tfrac{1}{2}R_{HBb}\sigma_{Bb}(2L_{DB}/d_B)^2 J_{1B}(V_{DB}, V_{XB})J_2(V_{DB}, V_{0B})$$

$$- R_{HGb}[(d_B + d_B)L_{DG}/R_N d_G^2]J_{1G}(V_{DG}, V_{XG})$$

$$A_{1,4} = -R_{HBb}\sigma_{Bb}d_B(2L_{DB}/d_B)^2 J_{1B}(V_{DB}, V_{0B})J_2(V_{DB}, V_{0B}),$$

[a] From Snejdar and Jerhot [196].
[b] Subscripts B and G on current densities refer to grain boundary and grain, respectively. V_{XG} and V_{XB} are the diffusion potentials at the edge of the space charge layers in the grain and grain boundary regions, respectively.

6. Minority Carrier Properties

Some attention has been given to the problems and mechanisms of recombination of minority carriers at the grain boundaries of polycrystalline thin films [197–204]. Much of the effort has been devoted to understanding the processes of minority carriers at grain boundaries in order to improve thin-film device performance and lifetime. Since a major concern involves thin-film solar cells, the behavior of minority carriers both in the dark and under illumination has been discussed.

A. GRAIN BOUNDARY MODEL AND DIFFUSION POTENTIAL

Card and Yang [197] systematically developed the dependence of the minority carrier lifetime τ on the doping concentration, grain size, and interface state density at the grain boundaries in polycrystalline Si. Other semiconductors have been likewise characterized, using the Card and Yang approach [197–202].

Figure 3.52 represents the band diagram of the region surrounding a grain boundary in an *n*-type semiconductor. For the dark case (Fig. 3.52a), the states are filled (in equilibrium) to the Fermi level, and the band

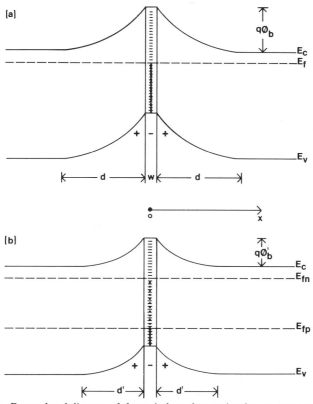

Fig. 3.52   Energy band diagram of the grain boundary region for an *n*-type semiconductor. (a) Dark case; (b) illuminated case. $q_{\phi b}$ and $q\phi'_b$ are the dark and light diffusion potentials, respectively (from Kazmerski [198]).

bending occurs in order to preserve charge neutrality. In the usual case, the width of the grain boundary (w) is much less than the width of the depleted region ($2d'$) thus balancing the charge:

$$Q_i = Q_d \qquad (3.101)$$

where Q_i is the net negative charge contained in the interface states and Q_d the net positive charge in the depletion region. Letting $E_v(0) = 0$, these charge densities are expressed [197, 198]

$$Q_c = qN_d(2d) = 8q\epsilon N_d(q\phi_b) \qquad (3.102)$$

and

$$Q_i = qN_i(2E_f/3q - E_v/q) \qquad (3.103)$$

where ϵ is the semiconductor permittivity, ϕ_b the diffusion potential, N_d the density of states (i.e., doping concentration), and N_i the grain boundary interface state density with units inverse square centimeters per electron volt. The diffusion potential ϕ_b can then be calculated by using these relationships in Eq. (3.101). A similar relationship can be obtained with a model for a p-type semiconductor. Figure 3.53 shows the dark diffusion potential for p-type CuInSe$_2$ as a function of the grain boundary interface density for various carrier concentrations. The interface state densities correspond roughly to the following grain boundary types:

$N_i > 10^{13}/(\text{cm}^2\,\text{eV})$:   High-angle grain boundaries;
$10^{11} < N_i < 10^{13}/(\text{cm}^2\,\text{eV})$:   Medium-angle grain boundaries;
$N_i < 10^{11}/(\text{cm}^2\,\text{eV})$:   Low-angle grain boundaries.

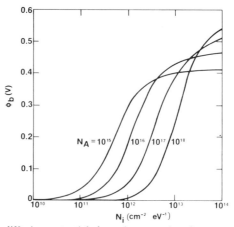

Fig. 3.53 Dark diffusion potential dependence on interface state density for various carrier concentrations in p-type CuInSe$_2$ (from Kazmerski [198]).

For a solar cell under illumination, the interface states adjust their charge by an initial net capture of holes (n-type semiconductor) or electrons (p-type). This, in turn, reduces the diffusion potential to a value $q\phi'_b$, which is $\sim 0.1q\phi_b$ and results in the maximum recombination rate. The grain boundary region under illumination is shown in Fig. 3.52b.

B. RECOMBINATION VELOCITY AND MINORITY CARRIER LIFETIME

In the case of solar cell material under illumination, the recombination velocity S of the minority carriers at the grain boundary can be estimated in a straightforward manner. The recombination current is defined [205] as

$$J_r = qS_0 p'_0 \simeq qS_0 p_0 \qquad (3.104)$$

where the subscript 0 indicates $x = 0$, p'_0 is the excess concentration, and p_0 the equilibrium concentration which can be expressed as $p_0 = p(d')$ $\exp(q\phi_b/kT)$ from Fig. 3.52b. If $p_0 \simeq n_0$, the recombination current is given by [206]

$$J_r = (q/2)N_i \sigma v (E_{fn} - E_{fp}) \qquad (3.105)$$

where σ is the capture cross section and v the thermal velocity of carriers. The recombination current at the edge of the depletion region has been calculated [197] to be

$$J_r \simeq qSp(d') \qquad (3.106)$$

And, combining Eq. (3.106) with Eqs. (3.104) and (3.105),

$$S \simeq \tfrac{1}{4}\sigma v (E_{fn} - E_{fp})N_i \exp(q\phi'_b/kT) \qquad (3.107)$$

Thus the diffusion potential ϕ'_b enhances the recombination at the grain boundary.

The dependence of S upon $q\phi'_b$ for various N_i is presented in Fig. 3.54 for CuInSe$_2$ ($\sigma = 2 \times 10^{-15}/\text{cm}^2$, $v = 10^7$ cm/s, $E_{fn} - E_{fp} \simeq 0.5$ eV). For $N_A \simeq 10^{16}/\text{cm}^3$, $q\phi'_b \simeq 0.13$ eV. Thus for high-angle grain boundaries ($N_i > 10^{13}/(\text{cm}^2\,\text{eV})$), recombination velocities in excess of 10^6 cm/s are expected.

The minority carrier lifetime can be calculated for a film with columnar-grain geometry. In this case, the volume recombination center density is [197]

$$N_r = 4N/l \qquad (3.108)$$

where N is the surface density at $x = d'$ [i.e., from examination of Eq. (3.107), $S = Nv$ or $N = \tfrac{1}{4}(E_{fn} - E_{fp})N_i \exp(q\phi'_b/kT)$]. Therefore,

$$N_r = (E_{fn} - E_{fp})(N_i/l)\exp(q\phi'_b/kT) \qquad (3.109)$$

Therefore, for a p-type semiconductor, the minority carrier lifetime is

$$\tau_n = 1/\sigma v N_r = l\exp(-q\phi'_b/kT)/\sigma v N_i(E_{fn} - E_{fp}) \qquad (3.110)$$

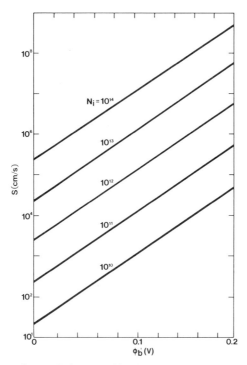

Fig. 3.54 Dependence of the recombination velocity upon the illuminated diffusion potential and interface state density in CuInSe$_2$ under 100 mW/cm^2 illumination (from Kazmerski [198]).

This dependence is shown in Fig. 3.55 for CuInSe$_2$. For the high-angle-grain boundary case, with typical 10^{-4} cm grain size, the minority carrier lifetime is in the 10^{-9}–10^{-10}-s range, which is more than one order of magnitude lower than that reported for the corresponding single crystal.

The validity of these calculations has been verified for the CuInSe$_2$–CdS thin-film solar cell [198, 200, 207] and by comparison to the cylindrical grain model (see Chapter 7). Rothwarf and Barnett [208] used a geometrical argument to account for grain bonndary recombination. Simply, all carriers generated closer to the grain boundary than the collecting site (i.e., the junction in the case of the solar cell) are lost. A good correspondence has been demonstrated [199] between the collection efficiency evolving from the minority carrier lifetime model [Eq. (3.110)] and that derived by Rothwarf and Barnett.

Leong and Yee [209] recently reported a standard photoconductance measurement for determining the intragrain recombination velocity of excess charge carriers in large-grain-size materials. This technique is similar to that described by Wang and Wallis [210] and simply monitors the

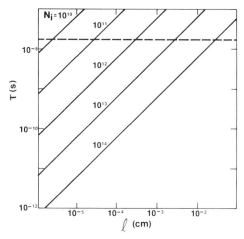

Fig. 3.55 Dependence of minority carrier lifetime on grain diameter for various inter-face state densities in $CuInSe_2$. Dashed line indicates single-crystal value of lifetime (from Kazmerski *et al.* [199]).

current passing perpendicular to the grain boundary. The analysis assumes that the incident light is absorbed entirely on the surface parallel to the grain boundary but in the grain itself. Thus no excess carriers are generated in the boundary region. Leong and Yee's analysis of the simple bicrystal case shows that (i) if the grain size l is much larger than the diffusion length L, the excess carrier lifetime in the polycrystalline material is not distinguishable from that in the comparable single crystal (i.e., having the same bulk and surface properties), and (ii) as the intergrain recombination velocity increases, the effective lifetime tends to saturate.

C.   MINORITY CARRIER MOBILITY IN POLYCRYSTALLINE THIN FILMS

Yee [211] has provided some insight into the behavior of minority carriers in polycrystalline semiconductors. He used the simple model shown in Fig. 3.56, similar to those of other investigators, in which the grain boundary is inverted with respect to the grain majority carrier type to

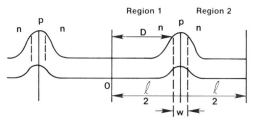

Fig. 3.56 Energy band model used in calculating minority carrier mobility for an *n*-type polycrystalline thin film (from Yee [211]).

estimate the minority carrier mobility. Considering the regions on each side of the grain boundary (labeled in Fig. 3.56), the carrier concentrations can be written for each:

Region 1: $\quad p(x) = p_{n0} + A_1 \exp(-x/L_p) + A_2 \exp(x/L_p)$ $\qquad$ (3.111)
$\qquad\qquad p(x) = p_{n0} \exp(q\psi_e/kT), \quad x = 1/2 - w/2$

with boundary conditions $p(x) = p_{n0}$, $x = 0$.

Region 2: $\quad p(x) = B_1 \exp(-x/L_p) + B_2 \exp(x/L_p) + p_{n0}$ $\qquad$ (3.112)
$\qquad\qquad p(x) = p_{n0} \exp(q\psi_c/kT), \quad x = 1/2 + w/2$

with boundary conditions $p(x) = p_{n0}$, $x = l$.

where the effective quasi-Fermi levels ψ_e and ψ_c are defined in the band diagram in Fig. 3.55. From Eqs. (3.110) and (3.111) and the boundary conditions, the diffusion equations are derived:

$$J_1 = -qD_p(\partial p/\partial x)|_{(1/2)-(w/2)}$$
$$\quad = -(qD_p p_{n0}/L_p)\left[\exp(q\psi_e/kT) - 1\right]\tanh(l/2L_p) \qquad (3.113)$$

and likewise

$$J_2 = (qD_p p_{n0}/L_p)\left[\exp(q\psi_c/kT) - 1\right]\tanh(l/2L_p) \qquad (3.114)$$

It can be observed from Fig. 3.57 that $\psi_e = -(V - \Delta V) = V - \psi_c$. It has been previously shown that [161]

$$\exp(q\,\Delta V/kT) = 2\left[1 + \exp(qV/kT)\right]^{-1} \qquad (3.115)$$

Therefore, the minority carrier current can now be derived by substituting Eq. (3.115) into either Eq. (3.113) or (3.114), making use of the relationships between the quasi-Fermi levels and ΔV and V. Thus,

$$J' = (qD_p p_{n0}/L_p)\tanh(qV/2kT)\tanh(l/2L_p) \qquad (3.116)$$

This can now be combined with the majority carrier current density derived by Seto [Eq. (3.80)] to give the total current. Assuming $qV/kT \ll 1$,

$$J = n_{av}q\left[ql/(2\pi m^* kT)^{1/2}\exp(-q\phi_b/kT)\right]$$
$$\quad + \bar{p}_n q\left[(\alpha l/2L_p)\mu\tanh(l/2L_p)\right] \qquad (3.117)$$

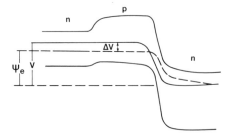

Fig. 3.57 Detailed band diagram in the vicinity of the grain boundary (from Yee [211]).

where $\bar{p}_n = (1/l)\int_0^l p(x)\,dx$, $\alpha = p_{n0}/\bar{p}$, and μ is the minority carrier mobility due to bulk and impurity contributions (i.e., $\mu^{-1} = \mu_b^{-1} + \mu_i^{-1}$). But since [203, 204]

$$J = qN_a\mu_n + q\bar{p}_n\mu_n \tag{3.118}$$

the mobility can be obtained using Eqs. (3.117) and (3.118):

$$\mu_n = \left[ql/(2\pi m^* kT)^{1/2}\right]\exp(-q\phi_b/kT) \tag{3.119}$$

$$\mu_n = (l/2L_p)\mu\tanh(l/2L_p) \tag{3.120}$$

Therefore the effective minority carrier mobility can be written [16] as

$$1/\mu_e = 1/\mu + 1/\mu_n \tag{3.121}$$

Or, solving for μ_e using Eqs. (3.120) and (3.121),

$$\mu_e = \mu\left[l/(l + \bar{L})\right] \tag{3.122}$$

where $\bar{L} = (2L_p)/[\alpha\tanh(l/2L_p)]$ and $L_p = [(kT/q)\mu\tau]^{1/2}$, in which τ is the lifetime associated with μ.

Yee extended this analysis to approximate the effects of grain size and interface (grain boundary) recombination velocity S on the minority carrier diffusion length. The effective diffusion length is defined as

$$L_{eff} = (kT\mu_e\tau_{eff}/q)^{1/2} \tag{3.123}$$

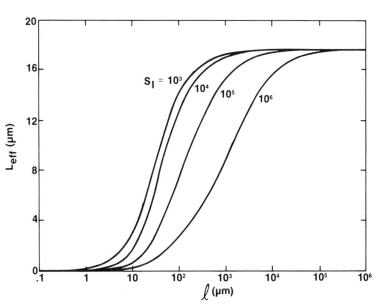

Fig. 3.58 Diffusion length of CdTe as a function of grain size for several grain boundary recombination velocities (from Yee [211]).

where the lifetime is given by the bulk, impurity, and the interface lifetimes τ_g'.

$$1/\tau_{eff} = 1/\tau + 1/\tau_g' \tag{3.124}$$

Since [195]

$$\tau_g \simeq (2l/3S)\exp(-q\phi_b/kT) \tag{3.125}$$

combining Eqs. (3.123) and (3.125) with the expression for the recombination velocity [Eq. (3.107)], the behavior of the effective diffusion length with grain size and S can be estimated. Figure 3.58 shows such an analysis for CdTe. Yee has shown some correlation with experimental data, as presented in Fig. 3.58.

7. Composite Models and Other Effects

A.   SURFACE AND GRAIN BOUNDARY SCATTERING

For semiconductor thin films, the carrier transport can be determined by more than one mechanism. The composite influence of the surface scattering mechanisms discussed in Section 3.1 and of the grain boundary scattering mechanism presented in Section 3.2 have been reported in the literature [7, 29, 56]. Considering the additive nature of noninteracting scattering mechanisms, the effective film mobility may be written

$$1/\mu_f = 1/\mu_g + 1/\mu_s \tag{3.126}$$

By incorporating Eqs. (3.28) and (3.59) into Eq. (3.126), one obtains [29]

$$\mu_f = \mu_b \exp(-q\phi_b/kT)(1 + 2\lambda/d)^{-1} \tag{3.127}$$

An examination of Eq. (3.127) as a function of film thickness shows that, for $d \gg \lambda$, the effect of surface scattering becomes negligible, and the mobility approaches a constant value since μ_b and μ_g do not depend upon thickness for a uniform film. However, if the thickness is on the order of magnitude of the mean surface scattering length λ, the mobility is reduced. This composite effect was already shown in Fig. 3.14 for CdS. From these data, λ can be calculated from the thickness-dependent portions of the curves, and $q\phi_b$ and μ_b can be evaluated from the constant-mobility, thicker-film portions.

Waxman et al. [32] developed a composite description of the behavior of polycrystalline films in their field-effect experiments (Section 3.1). The effective barrier height $q\phi'$ of the grain boundaries was calculated for a change in the carrier density and a surface depletion region of width L_c due to the band bending. Thus [32]

$$q\phi' = kT \ln[(n_1 + \Delta n/L_c)/(n_2 + \Delta n/L_c)] \tag{3.128}$$

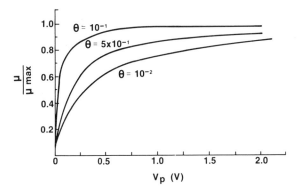

Fig. 3.59 Calculated dependence of Hall mobility, normalized to maximum value, on surface potential for various surface state densities. θ_s is the ratio of induced free charge to trapped charge in surface states and traps (from Waxman *et al.* [32]).

where n_1 and n_2 are the original carrier densities in the grain and grain boundary, respectively [see Eq. (3.69)], and Δn the induced concentration. From this, the change in barrier potential can be calculated to be

$$q\phi = q\phi - q\phi' = -kT \ln\left[(1 + \alpha V_p)/(1 + \beta V_p)\right] \qquad (3.129)$$

where

$$\alpha = \theta\epsilon_i/qd_iL_cn_1, \qquad \beta = \theta\epsilon_i/qd_iL_cn_2$$

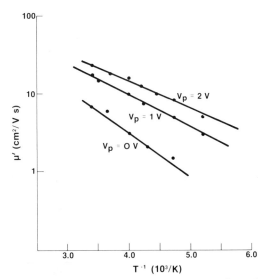

Fig. 3.60 Variation in Hall mobility with inverse temperature for various applied surface potentials in SiO–CdS thin-film field-effect structure (from Waxman *et al.* [32]).

and ϵ_i is the permittivity of the insulator in the field-effect structure, d_i the insulator film thickness, and θ the ratio of induced free charge to trapped charge in surface states and traps $(= q\Delta nd_i/\epsilon_i V_p)$ [32]. The film mobbility becomes

$$\mu_f = (C/kT)\left[(1 + \beta v_p)/(1 + \alpha V_p)\right]\exp(-q\phi/kT) \qquad (3.130)$$

where $C = qv/4n_1$. The effective film mobility is predicted to rise as a function of V_p, as shown in Fig. 3.59 for various values of θ. The composite effect on inverse temperature is presented in Fig. 3.60, in which the field-injected excess charge reduces the barrier height by counteracting the depletion of charge at the grain boundary barriers.

B.   OTHER CONTRIBUTIONS

Dislocations

Semiconductor thin films tend to have large concentrations of defects due to growth processes [1, 212], to lattice mismatches between layers [213–216], to differences in thermal expansion coefficients between adjacent films or substrate and film [214], and to stresses [215–220]. Surface defects on the substrate or underlayer film can also extend into subsequently deposited layers. The most common among these defects is the dislocation [212]. Overviews of the relationship between dislocations and the electrical characteristics of semiconductors are reported by Broudy [221], Read [222], and Matare [182].

Dislocations, like grain boundaries, provide charged deformation regions for carrier scattering. The mobility dependence upon temperature can be written [223]

$$1/\mu_d \simeq C_1/T^{3/2} + C_2 T \qquad (3.131)$$

where the first term corresponds to charged-imperfection scattering and the second term to deformation scattering of the carriers, with C_1 and C_2 temperature-independent constants. Broudy [221] and Dexter and Seitz [223] derived the mobility contributions due to scattering from a uniform distribution of noncharged dislocations:

$$\mu_d = \gamma N_d/kT \qquad (3.132)$$

where $\gamma = (3\pi/32)(E^2 b^2 m^*/hq)[(1 - 2\nu)/(1 - \nu)]^2$, b is the dislocation Burger's vector, ν Poisson's ratio, and E the deformation potential. The effects of dislocations on the electrical properties become more important in thicker, single-crystal films in which grain boundary and surface scattering mechanisms do not dominate the transport.

Stacking Faults

Another intragrain defect is the stacking fault, a planar defect which results from the multiple nucleation and subsequent growth to continuity for thin films [224]. A number of investigations of the electrical properties of these defects have been published [225–231]. Brown [225] treats the scattering of plane wave electrons from the stacking faults, which are represented as constant-potential surfaces. He derives an expression for the resistivity of such defects in semiconductors:

$$\rho_{sf} = \beta\left(6\pi h^3 / q^2 m^* E\right)\bar{R} \tag{3.133}$$

where β is the linear fault density, $E = h^2 k^2 / 2m^*$, the average electron energy, and $\bar{R}$ the average value of the reflection coefficient for all faults.

The effects of stacking faults on electronic transport have also been modeled similarly to the grain boundary; that is, for a sufficiently faulted film [229, 230],

$$\rho = \rho_b \exp(q\phi_{sf}/kT) \tag{3.134}$$

After the resistivity is related to the mobility and an anisotropy in the carrier concentration is assumed, the film mobility can be expressed [230] as

$$\mu_f = \mu_b\left[\exp(-q\phi_b/kT)(1 + 2\lambda/d)^{-1}\right]\left\{1 + m\exp\left[(q\phi_{sf} - q\phi_b)/kT\right]\right\}^{-1} \tag{3.135}$$

where m is a stacking fault count density (faults per grain) and $q\phi_{sf}$ the stacking fault potential barrier. If $m = 0$, Eq. (3.134) reduces to the grain boundary-dominated model given by Eq. (3.127). It should be noted that the value of m depends geometrically on the relative orientation of the stacking faults and the grain boundaries [230]. The validity of Eq. (3.134) has been demonstrated for highly faulted CdS films that have some degree of grain misorientation. For these films, the fault densities were found to be 10^3–10^6/cm, with 5–15 faults per grain. The magnitude of $q\phi_{sf}$ was estimated to be 0.035 eV. A substrate temperature dependence of the fault density was reported with highest densities, corresponding to minimum film mobilities, occurring at $T_{sub} = 220°C$ for glass substrates [230, 231].

Note added in proof: Recent work by Ghosh, Fishman, and Feng (*J. Appl. Phys.* **51**, 446 (1980)) presents a discussion of the role of grain boundaries as traps and recombination centers in polycrystalline Si. Agreement between theory and experiment for the effects of such grain boundaries in polycrystalline Si solar cells for almost all device parameters is reported.

ACKNOWLEDGMENTS

The author wishes to acknowledge and express his appreciation to Dr. Sigurd Wagner who provided both assistance and useful suggestions in his reviews of this chapter. Sincere

128 LAWRENCE L. KAZMERSKI

appreciation and recognition are given to Susan Sczepanski who patiently provided the numerous figures for this chapter, and to Bobbie Fry and Betsy Fay-Saxon who prepared the manuscript through its many revisions. The author also thanks P. J. Ireland for his many helpful readings of this chapter. Finally, the author is indebted to Joyce Barrett who gave considerable support through her worthwhile and needed editorial suggestions.

REFERENCES

1. See Appendix I of this book for a list of relevant texts on thin-film properties.
2. J. E. Morris and T. J. Coutts, *Thin Solid Films*, **47**, 3 (1977).
3. L. I. Maissel, "Handbook of Thin Film Technology" (L. I. Maissel, ed.), Chapter 13. McGraw-Hill, New York, 1970.
4. See, for example, T. J. Coutts, "Active and Passive Thin Film Devices" (T. J. Coutts, ed.), Chapter 3. Academic Press, New York, 1978.
5. J. G. Simmons, "Handbook of Thin Film Technology" (L. I. Maissel, ed.), Chapter 14. McGraw-Hill, New York, 1970.
6. J. C. Anderson, *Adv. Phys.* **19**, 311 (1970).
7. R. H. Bube, *Ann. Rev. Mater. Sci.* **5**, 201 (1975).
8. R. F. Greene, *J. Phys. Chem. Solids* **14**, 291 (1960).
9. R. F. Greene, *Surf. Sci.* **2**, 101 (1964).
10. R. F. Greene, *Phys. Rev.* **141**, 687 (1966).
11. B. Tavger and W. Kogan, *Phys. Lett.* **19**, 353 (1965).
12. B. Tavger and E. Erukhimov, *Zh. Eksper. Teor. Fiz.* **51**, 528 (1966).
13. B. Tavger, *Phys. Status. Solidi* **22**, 31 (1967).
14. See, for example, A. Many, Y. Goldstein, and N. B. Grover, "Semiconductor Surfaces," Chapter 2. North-Holland Publ., Amsterdam, 1965.
15. See, for example, J. L. Moll, "Physics of Semiconductors," Chapters 2 and 3. McGraw-Hill, New York, 1964.
16. A. Matthiessen, *Rep. Br. Assoc.* **32**, 144 (1862).
17. H. F. Wolf, "Semiconductors," pp. 281–283. Wiley (Interscience), New York, 1971.
18. K. Fuchs, *Proc. Cambridge Phil. Soc.* **11**, 120 (1938).
19. E. H. Sondheimer, *Adv. Phys.* **1**, 1 (1952).
20. M. S. P. Lucas, *J. Appl. Phys.* **36**, 1632 (1965).
21. J. N. Zemel, *Phys. Rev.* **112**, 762 (1958).
22. D. R. Frankl, "Electrical Properties of Semiconductor Surfaces," pp. 42–45. Pergamon, Oxford, 1967.
23. H. Fleitmer, *Phys. Status. Solidi* **1**, 483 (1961).
24. D. S. Campbell, "The Use of Thin Films in Physical Investigations" (J. C. Anderson, ed.), p. 299. Academic Press, New York, 1966.
25. H. Mayer, "Structure and Properties of Thin Films," p. 225. Wiley, New York, 1959.
26. K. L. Chopra, "Thin Film Phenomena," Chapters 6 and 7. McGraw-Hill, New York, 1969.
27. A. Many, Y. Goldstein, and N. B. Grover, "Semiconductor Surfaces," Chapters 4 and 8. North-Holland Publ., Amsterdam, 1965.
28. F. F. Ham and D. C. Mattis, *IBM J. Res. Dev.* **4**, 143 (1960).
29. L. L. Kazmerski, W. B. Berry, and C. W. Allen, *J. Appl. Phys.* **43**, 3515 (1972).
30. L. L. Kazmerski, *Thin Solid Films* **21**, 273 (1974).
31. H. F. Wolf, "Semiconductors," p. 287. Wiley (Interscience), New York, 1971.
32. A. Waxman, V. E. Henrich, F. V. Shallcross, H. Borkan, and P. K. Weimer, *J. Appl. Phys.* **36**, 168 (1965).

33. D. W. Covington and D. C. Ray, *J. Appl. Phys.* **45**, 2616 (1974).
34. A. V. Sachenko, *Sov. Phys.—Semicond.* **11**, 264 (1977). Also, B. D. Kandilarov and V. Detcheva, *J. Phys. C—Solid State* **10**, 1703 (1977).
35. C. Juhasz, Ph.D. Thesis, Univ. of London (1968). Also, C. Juhasz and J. C. Anderson, *Radio Electron. Eng.* **33**, 223 (1967).
36. R. Hezel, *Siemens Forsch.—V. Entwickl* **3**, 160 (1974).
37. A. Amith, *J. Phys. Chem. Solids* **14**, 271 (1960).
38. J. N. Zemel and J. O. Varela, *J. Phys. Chem. Solids* **14**, 142 (1960).
39. J. N. Zemel, J. D. Jensen, and R. B. Schoolar, *Phys. Rev.* **140**, A330 (1965).
40. J. N. Zemel and R. L. Petritz, *J. Phys. Chem. Solids* **8**, 102 (1959).
41. R. L. Petritz, *Phys. Rev.* **110**, 1254 (1958).
42. J. R. Schreiffer, *Phys. Rev.* **97**, 641 (1955).
43. H. H. Weider, *Thin Solid Films* **31**, 123 (1976).
44. A. Y. Cho and J. R. Arthur, *Prog. Solid-State Chem.* **10**, 157 (1975).
45. A. Y. Cho, *J. Vac. Sci. Technol.* **13**, 275 (1979).
46. J. R. Arthur, *J. Vac. Sci. Technol.* **13**, 273 (1979).
47. J. E. Davy, R. G. Turner, T. Pankey, and M. D. Montgomery, *Solid State Electron.* **6**, 205 (1963).
48. B. W. Sloope and C. O. Tiller, *J. Appl. Phys.* **38**, 140 (1967).
49. J. E. Davy, *Appl. Phys. Lett.* **8**, 164 (1966).
50. R. L. Ramey and W. D. McLennan, *J. Appl. Phys.* **38**, 3491 (1967).
51. C. C. Mai, T. S. Whitehouse, R. C. Thomas, and D. R. Goldstein, *J. Electrochem. Soc.* **118**, 331 (1971).
52. J. N. Zemel and M. Kaplit, *Surf. Sci.* **13**, 17 (1969).
53. M. Hamasaki, T. Adachi, S. Wakayama, and M. Kikuchi, *Solid State Commun.* **21**, 591 (1977).
54. D. J. Dumin and P. H. Robinson, *J. Appl. Phys.* **39**, 2759 (1968).
55. M. H. Brodsky and J. N. Zemel, *Phys. Rev.* **155**, 780 (1967).
56. W. B. Berry and T. S. Jayadevaiah, *Thin Solid Films* **3**, 77 (1969).
57. B. V. Izvochikov and I. A. Taksami, *Sov. Phys.—Semicond.* **1**, 985 (1968).
58. C. A. Neugebauer, *J. Appl. Phys.* **39**, 3177 (1968).
59. R. K. Swank, *Phys. Rev.* **153**, 844 (1967).
60. W. H. Leighton, *J. Appl. Phys.* **44**, 5011 (1973).
61. A. Amith, *J. Vac. Sci. Technol.* **15**, 353 (1978).
62. H. Berger, W. Kahle, and G. Janiche, *Phys. Status. Solidi* **28**, K97 (1968).
63. K. V. Shalimova, A. F. Andrusko, V. A. Dmitriev, and L. P. Pavlov, *Izv. Vuzov (Mos.)* **3**, 134 (1964).
64. R. W. Glew, *Thin Solid Films* **52**, 59 (1977).
65. See, for example, F. V. Wald, *Rev. Phys. Appl.* **12**, 277 (1977).
66. A. C. Thorsen, H. M. Manasevit and R. H. Harada, *Solid-State Electron.* **17**, 855 (1974).
67. G. E. Stillman and C. M. Wolf, *Thin Solid Films* **31**, 69 (1976).
68. Y. F. Orgin, V. N. Lutskii, and M. I. Elinson, *Sov. Phys. JETP Lett.* **3**, 71 (1966).
69. Y. F. Komnik and E. I. Bukhstab, *Sov. Phys. JETP Lett.* **6**, 58 (1967).
70. V. P. Duggal, R. Rup, and P. Tripathi, *Appl. Phys. Lett.* **9**, 293 (1966).
71. K. L. Chopra, L. C. Bobb, and M. H. Francombe, *J. Appl. Phys.* **34**, 1699 (1963).
72. Y. F. Komnik, E. I. Bukhshtab, and Y. V. Nikitin, *Thin Solid Films* **52**, 361 (1978).
73. J. G. Dash, "Films on Solid Surfaces," Chapters 1–5. Academic Press, New York, 1975.
74. G. Wedler, "Chemisorption." Butterworths, London, 1976.
75. S. R. Morrison, *Adv. Catal.* **7**, 259 (1955).

76. S. Baidyaroy and P. Mark, *Surf. Sci.* **30**, 53 (1972).
77. G. South and D. M. Hughes, *Thin Solid Films* **20**, 135 (1974).
78. C. C. Chen, A. H. Clark, and L. L. Kazmerski, *Thin Solid Films* **32**, L5 (1976).
79. T. J. Coutts, "Active and Passive Thin Film Devices" (T. J. Coutts, ed.), Chapter 5. Academic Press, New York, 1978.
80. H. F. VanHeek, *Solid-State Electron.* **11**, 459 (1968).
81. A. C. Ipri, *J. Appl. Phys.* **43**, 2770 (1972).
82. C. A. Neugebauer and R. E. Joynson, Abstr. 13, *Nat. Vac. Symp.* **13**, 105 (1967).
83. See, for example, J. C. Anderson, "Active and Passive Thin Film Devices" (T. J. Coutts, ed.), Chapter 6. Academic Press, New York, 1978.
84. T. L. Chu, *J. Cryst. Growth* **39**, 45 (1977). Also, T. L. Chu, S. S. Chu, G. A. VanderLeen, C. J. Lin, and J. R. Boyd, *Solid-State Electron.* **21**, 781 (1978).
85. Dieter Helmreich, Wacker Heliotronic GmbH (personal communication).
86. R. F. Egerton, Ph.D. Thesis, Univ. of London (1969).
87. J. Volger, *Phys. Rev.* **9**, 1023 (1950).
88. R. L. Petritz, *Phys. Rev.* **104**, 1508 (1956).
89. H. Berger, *Phys. Status Solidi* **1**, 739 (1961).
90. R. G. Mankarious, *Solid-State Electron.* **7**, 702 (1964).
91. L. L. Kazmerski, M. S. Ayyagari, and G. A. Sanborn, *J. Appl. Phys.* **46**, 4685 (1975).
92. L. L. Kazmerski, M. S. Ayyagari, F. R. White, and G. A. Sanborn, *J. Vac. Sci. Technol.* **13**, 139 (1976).
93. L. L. Kazmerski and Y. J. Juang, *J. Vac. Sci. Technol.* **14**, 769 (1977).
94. L. L. Kazmerski, F. R. White, M. S. Ayyagari, Y. J. Juang, and R. P. Patterson, *J. Vac. Sci. Technol.* **14**, 65 (1977).
95. R. Kassing and W. Bax, *Jpn. J. Appl. Phys. Suppl.* **2**, 801 (1974).
96. Z. T. Kuznicki, *Thin Solid Films* **33**, 349 (1976).
97. Z. T. Kuznicki, *Solid-State Electron.* **19**, 894 (1976).
98. V. Snejdar, *Slaboproudy Ob.* **31**, 293 (1970).
99. J. L. Davis and R. F. Greene, *Appl. Phys. Lett.* **11**, 37 (1967).
100. D. P. Snowden and A. M. Portis, *Phys. Rev.* **120**, 1983 (1960).
101. R. H. Bube, *Appl. Phys. Lett.* **13**, 136 (1968).
102. K. Kipskis, A. Sakalas and J. Viscakas, *Phys. Status. Solidi (a)* **4**, K217 (1971).
103. G. H. Blout, R. H. Bube, and A. L. Robinson, *J. Appl. Phys.* **41**, 2190 (1970).
104. S. S. Minn, *J. Res. Centre Nat. Res. Sci. Lab. Bellevue* **51**, 131 (1960).
105. C. Wu and R. H. Bube, *J. Appl. Phys.* **45**, 648 (1974).
106. J. Dresner and F. V. Shallcross, *Solid-State Electron.* **5**, 205 (1962).
107. N. F. Foster, *Proc. IEEE* **53**, 1400 (1965).
108. F. V. Shallcross, *Trans. AIME* **236**, 309 (1966).
109. R. W. Muller and B. G. Watkins, *Proc. IEEE* **52**, 425 (1964).
110. J. I. B. Wilson and J. Woods, *J. Phys. Chem. Solids* **34**, 171 (1973).
111. D. W. Readey, *J. Am. Ceram. Soc.* **49**, 681 (1966).
112. H. G. Dill and R. Zuleeg, *Solid-State Technol.* **1**, 27 (1964).
113. J. Dresner and F. V. Shallcross, *J. Appl. Phys.* **34**, 2390 (1963).
114. G. Hecht, J. Herberger, and C. Weissmantel, *Thin Solid Films* **2**, 293 (1968).
115. For a comprehensive review, see A. G. Stanley, Cadmium Sulfide, Vols. 1 and 2. Jet Propulsion Laboratory, Publ. 78-77; Pasadena, California, 1978.
116. K. Shimizu, *Jpn. J. Appl. Phys.* **4**, 627 (1965).
117. D. E. Brodie and J. LaCombe, *Can. J. Phys.* **45**, 1353 (1967).
118. R. Glang, J. G. Kren, and W. J. Patrik, *J. Electrochem. Soc.* **110**, 407 (1963).
119. K. Mitchell, A. L. Fahrenbruck, and R. H. Bube, *J. Vac. Sci. Technol.* **12**, 909 (1975).
170. F. H. Nicoll, *J. Electrochem. Soc.* **110**, 1165 (1963).

121. D. A. Cusano, "Physics and Chemistry of II-VI Compounds" (M. Aven and J. S. Prener, eds.), Chapter 14. Wiley, New York, 1967.
122. S. Espevik, C. Wu, and R. H. Bube, *J. Appl. Phys.* **42**, 3513 (1971).
123. L. S. Palatnik and V. K. Sorokin, *Isv. Vyssh. Uch. Zaved.* **3**, 48 (1965).
124. H. Gobrecht, K. W. Boeters, and H. J. Fleisher, *Z. Phys.* **187**, 232 (1965).
125. L. S. Palatnik and V. K. Sorokin, *Sov. Phys.-Solid State* **8**, 869 (1966).
126. Y. Makino, *Jpn. J. Phys. Soc.* **19**, 580 (1964).
127. I. P. Voronia and S. A. Semiletov, *Soc. Phys.-Solid State* **6**, 1494 (1964).
128. D. L. Lile and J. C. Anderson, *J. Phys. D.* **2**, 839 (1969).
129. C. Paparoditis, *J. Phys.* **25**, 226 (1964).
130. H. M. Manasevit and W. I. Simpson, *J. Electrochem. Soc.* **120**, 135 (1975).
131. M. J. Tsai and R. H. Bube, *J. Appl. Phys.* **49**, 3397 (1978).
132. J. H. McFee, B. I. Miller, and K. J. Bachmann, *J. Electrochem. Soc.* **124**, 259 (1977).
133. K. J. Bachmann, E. Buehler, J. L. Shay, S. Wagner, and M. Bettini, *J. Electrochem. Soc.* **123**, 109 (1976).
134. R. F. Potter and H. H. Wieder, *Solid-State Electron.* **7**, 153 (1964).
135. C. Juhasz and J. C. Anderson, *Phys. Lett.* **12**, 163 (1964).
136. W. J. Williamson, *Solid-State Electron.* **9**, 213 (1966).
137. L. R. Shiozawa, F. Augustine, G. A. Sullivan, J. M. Smith, and W. R. Cook, Clevite Corp. Final Rep., AF33(615)-5224 (1969).
138. S. G. Ellis, *J. Appl. Phys.* **38**, 2906 (1967).
139. S. R. Das, R. Nath, A. Banerjee, and K. L. Chopra, *Solid-State Commun.* **21**, 49 (1971).
140. H. Nimura, A. Atoda, and T. Nakau, *Jpn. J. Appl. Phys.* **16**, 403 (1977).
141. L. L. Kazmerski and D. M. Racine, *Thin Solid Films* **30**, L19 (1975).
142. H. R. Riedl, R. B. Schoolar and B. Houston, *Solid-State Commun.* **4**, 399 (1966).
143. J. R. Knight, D. Effer, and P. R. Evans, *Solid-State Electron.* **8**, 178 (1965).
144. H. Poth, H. Bruch, M. Heyer, and P. Balk, *J. Appl. Phys.* **49**, 285 (1978).
145. W. V. McLevige, K. V. Vaidyanathan, and B. G. Streetman, *Appl. Phys. Lett.* **33**, 127 (1978).
146. H. Poth, *Solid-State Electron.* **21**, 801 (1978).
147. A. E. Blakeslee and S. M. Vernon, *IBM J. Res. Dev.* **22**, 346 (1978).
148. G. S. Kamath and D. Bowman, *J. Electrochem. Soc.* **114**, 192 (1967).
149. L. L. Kazmerski, "Ternary Compounds 1977" (G. D. Holah, ed.), pp. 217–228. Inst. of Phys. Conf. Series, London, 1977.
150. Y. Kokubun and M. Wada, *Jpn. J. Appl. Phys.* **16**, 879 (1977).
151. E. Elliott, R. D. Tomlinson, J. Parkes, and M. J. Hampshire, *Thin Solid Films* **20**, S25 (1974).
152. See, for example, "Direct Solar Energy Conversion for Terrestial Use" (K. W. Boer, ed.). Final Report, NSF/RANN/AER72-03478 A03/FR/75, 1975.
153. G. H. Blout, R. H. Bube, and A. L. Robinson, *J. Appl. Phys.* **41**, 2190 (1970).
154. K. Tanaka and Z. Huruhata, *J. Electrochem. Soc. Jpn.* **37**, 133 (1969).
155. M. E. Crowder and T. O. Sedgewick, *J. Electrochem. Soc.* **119**, 1565 (1972).
156. A. L. Fripp, *J. Appl. Phys.* **46**, 1240 (1975).
157. T. I. Kamins, *J. Appl. Phys.* **42**, 4357 (1971).
158. P. Rai-Choudhury and P. L. Hower, *J. Electrochem. Soc.* **120**, 1761 (1973).
159. J. Y. W. Seto, *J. Electrochem. Soc.* **122**, 701 (1975).
160. W. E. Taylor, N. H. Odell, and H. V. Fan, *Phys. Rev.* **88**, 867 (1952).
161. R. K. Mueller, *J. Appl. Phys.* **32**, 635 (1961).
162. R. K. Mueller, *J. Appl. Phys.* **32**, 640 (1961).
163. J. Y. W. Seto, *J. Appl. Phys.* **46**, 5247 (1975).
164. G. Baccarani, B. Ricco, and G. Spandini, *J. Appl. Phys.* **49**, 5565 (1978).

165. C. H. Seager and T. G. Castner, *J. Appl. Phys.* **49**, 3879 (1978).
166. G. Baccarani, G. Masetti, M. Severi, and G. Spandini, *Proc. Int. Symp. Silicon Mat. Sci. Technol., 3rd, Philadelphia, Pennsylvania* (1977).
167. T. I. Kamims, *IEEE Trans. Parts Hyb. Packag.* **VHP-10**, 221 (1974).
168. J. Manoliu and T. I. Kamins, *Solid-State Electron.* **15**, 1103 (1972).
169. See, for example, A. Many, Y. Goldstein, and N. B. Grover, "Semiconductor Surfaces," Chapters 5 and 9. North-Holland Publ., Amsterdam, 1965.
170. See, for example, "The Physics of SiO_2 and Its Interfaces " (S. T. Pantelides, ed.). Pergamon, Oxford, 1978.
171. J. R. Monkowski, J. Bloem, L. J. Giling, and M. W. M. Graef, *Appl. Phys. Lett.* **35**, 410 (1979).
172. C. H. Seager and T. G. Castner, *Bull. Am. Phys. Soc.* **22**, 434 (1977).
173. J. W. Cleland, R. D. Westbrook, R. F. Wood, and R. T. Young, *Proc. Nat. Wksp. Low-Cost Polycryst. Silicon Solar Cells, Dallas* pp. 113–117 (1977).
174. C. H. Seager and D. S. Ginley, *App. Phys. Lett.* **34**, 337 (1979).
175. G. E. Pike and C. H. Seager, *J. Appl. Phys.* **50**, 3414 (1979).
176. C. H. Seager, G. E. Pike, and D. S. Ginley, *Phys. Rev. Lett.* **43**, 532 (1979).
177. C. H. Seager and D. S. Ginley, *Proc. Photovoltaic Mat. Device Measurements Workshop: Focus Polycryst. Thin Film Cells—Virginia* pp. 115–120. Solar Energy Research Institute, Golden, Colorado, 1979.
178. C. H. Seager and G. E. Pike, *Appl. Phys. Lett.* **35**, 709 (1979).
179. L. L. Kazmerski and P. J. Ireland, *Proc. Photovoltaic Mat. Device Measurements Workshop: Focus Polycryst. Thin Film Cells—Virginia* pp. 145–149. Solar Energy Research Institute, Golden, Colorado, 1979.
180. L. L. Kazmerski and P. J. Ireland, *J. Vac. Sci. Technol.* **17**, (1980).
181. L. L. Kazmerski, P. J. Ireland, and T. F. Ciszek, *Appl. Phys. Lett.* **36**, 323 (1980).
182. H. F. Matare, "Defect Electronics in Semiconductors." Wiley (Interscience), New York, 1971.
183. F. L. Vogel, W. T. Read, and L. C. Lovell, *Phys. Rev.* **94**, 1791 (1954).
184. J. P. McKelvey, *Phys. Rev.* **106**, 910 (1957).
185. E. I. Goldman, I. B. Gulyaev, A. G. Zhdan, V. B. Sandomirskii, and V. P. Khrenov, *Sov. Phys.—Semicond.* **9**, 905 (1976).
186. R. A. Brown, *J. Phys. F* **7**, 1477 (1977).
187. P. Guyot, *Phys. Status. Solidi* **38**, 409 (1970).
188. G. Landwehr and P. Handler, *J. Phys. Chem. Solids* **23**, 891 (1962).
189. Y. Makukura, *J. Phys. Soc. Jpn.* **16**, 842 (1961).
190. P. D. Maycock, *Proc. IEEE Photovoltaics Spec. Conf., 13th, Washington, D.C.*, pp. 5–8. IEEE, New York, 1978.
191. J. C. Anderson, *Thin Solid Films* **50**, 25 (1978). Also, H. Koelmans, *ibid.* **8**, 19 (1971).
192. M. J. Cohen, J. S. Harris, and J. R. Waldrop, *Proc. GaAs Conf., St. Louis. Inst. Phys. Conf. Ser.* **45**, 263 (1979).
193. See, for example, L. E. Davis, N. C. MacDonald, P. W. Palmberg, G. E. Riach, and R.E. Weber, "Handbook of Auger Electron Spectroscopy." Physical Electronics Industries, Eden Prairie, Minnesota, 1976.
194. P. D. Dapkus *et al.*, *Proc. IEEE Photovoltaics Spec. Conf., 13th, Washington, D.C.* pp. 960–965. IEEE, New York, 1979.
195. J. Jerhot and V. Snejdar, *Thin Solid Films* **52**, 379 (1978).
196. V. Snejdar and J. Jerhot, *Thin Solid Films*, **37**, 303 (1976).
197. H. C. Card and E. S. Yang, *IEEE Trans. Electron Dev.* **ED-24**, 397 (1977).
198. L. L. Kazmerski, *Solid-State Electron.* **21**, 1545 (1978).

199. L. L. Kazmerski, P. Sheldon, and P. J. Ireland, *Thin-Solid Films* **58**, 95 (1979).
200. L. L. Kazmerski, P. Sheldon, and P. J. Ireland, *Am. Vac. Soc. Symp., 25th*, San Francisco, November 1978.
201. P. T. Landsberg and C. M. Kimpke, *Proc. IEEE Photovoltaics Spec. Conf., 13th, Washington, D.C.* pp. 665–666. IEEE, New York, 1978.
202. R. Singh, T. N. Bhar, J. Shewchun, and J. J. Loferski, *J. Vac. Sci. Technol.* **13**, 236 (1979).
203. L. M. Fraas, *J. Appl. Phys.* **49**, 871 (1978).
204. C. Feldman, N. A. Blum, H. K. Charles, and F. G. Satkiewicz, *J. Electron. Mat.* **7**, 309 (1978).
205. S. M. Sze, "Physics of Semiconductor Devices," pp. 22–149, 640–653. Wiley, New York, 1969.
206. A. G. Milnes and D. L. Feucht, "Heterojunctions and Metal-Semiconductor Junctions," pp. 3–29, 51–57, 125–142. Academic Press, New York, 1972.
207. L. L. Kazmerski, P. J. Ireland, F. R. White, and R. B. Cooper, *Proc. IEEE Photovoltaics Spec. Conf., 13th, Washington, D.C.* pp. 184–189. IEEE, New York, 1978.
208. A. Rothwarf and A. M. Barnett, *IEEE Trans. Electron Dev.* **ED-24**, 381 (1977).
209. J. Y. Leong and J. H. Yee, *Appl. Phys. Lett.* **35**, 601 (1979).
210. S. Wang and G. Wallis, *Phys. Rev.* **105**, 1459 (1957).
211. J. H. Yee, The Systematic Computation of the Performance of Photovoltaic Cells Based on First Principles. DOE Rep. W-7405-ENG-48 (Modeling) (1979).
212. See, for example, S. Amelinckx, "The Direct Observation of Dislocations." Academic Press, New York, 1964.
213. J. L. Shay, S. Wagner, and J. C. Phillips, *Appl. Phys. Lett.* **28**, 31 (1976).
214. S. Wagner, *J. Cryst. Growth* **31**, 113 (1975).
215. Z. I. Alferov, *Soc. Phys.—Semicond.* **11**, 1216 (1978).
216. K. A. Jones, C. H. Chang, and B. F. Shirreffs, *Proc. IEEE Photovoltaics Spec. Conf., 13th, Washington, D.C.* pp. 513–518. IEEE, New York, 1978.
217. R. A. Logan, G. L. Pearson, and D. A. Kleinman, *J. Appl. Phys.* **30**, 855 (1959).
218. P. Chaudhari, *J. Vac. Sci. Technol.* **9**, 520 (1972).
219. O. Simpson, *Nature (London)* **160**, 791 (1947).
220. R. W. Hoffman, *Thin Solid Films*, **34**, 185 (1976).
221. R. M. Broudy, *Adv. Phys.* **12**, 135 (1963).
222. W. T. Read, *Phil. Mag.* **45**, 775 (1954).
223. D. L. Dexter and F. Seitz, *Phys. Rev.* **86**, 964 (1952).
224. See, for example, W. Bollman, "Crystal Defects and Crystalline Interfaces," pp. 78–97. Springer-Verlag, Berlin and New York, 1970.
225. R. A. Brown, *Phys. Rev.* **156**, 692 (1967).
226. A. Howie, *Phil. Mag.* **5**, 251 (1960).
227. H. Blank, P. Delavignette, and S. Amelinckx, *Phys. Status. Solidi* **2**, 1660 (1964).
228. R. H. Bube, "Photoconductivity of Solids," pp. 417–420. Wiley, New York, 1960.
229. L. L. Kazmerski, N. B. Berry, and C. W. Allen, *Proc. Nat. Electron. Conf.* **26**, 202 (1970).
230. L. L. Kazmerski, W. B. Berry, and C. W. Allen, *J. Appl. Phys.* **43**, 3521 (1972).
231. L. L. Kazmerski, W. B. Berry, and C. W. Allen, "Solar Cells," pp. 141–154. Gordon and Breach, New York, 1971.

4 Optical Properties of Polycrystalline Semiconductor Films

A. H. CLARK

University of Maine at Orono
Orono, Maine

4.1 INTRODUCTION

Semiconductor thin films have been produced and studied in polycrystalline form for many decades. Indeed, most new semiconducting materials are produced in polycrystalline form before techniques are developed for producing single crystals. Thus the literature contains hundreds of papers detailing optical properties of polycrystalline semiconductor films. In the vast majority of these papers, the effects of grain boundaries are neglected on the assumption (usually justified) that such effects on the optical properties are small.

POLYCRYSTALLINE AND AMORPHOUS
THIN FILMS AND DEVICES

With the need, however, to produce low-cost, large-area semiconducting devices, particularly photovoltaic devices, one is often faced with optimization of a polycrystalline material as an end product rather than an intermediate step. It then becomes important to understand the role of grain boundaries in determining electrical and optical properties.

In this review we shall for the most part restrict our discussion of optical properties to optical constants related to electronic structure in the material. Thus we will omit several topics which might come under the heading "optical properties." Such topics include photoconductivity (discussed in several other chapters of this book), luminescence, photoemission, infrared properties determined by lattice vibrations, and a variety of inelastic scattering phenomena. Undoubtedly, as polycrystalline films become increasingly important for practical applications, a wider range of optical properties will be studied in an attempt to understand the role of grain boundaries. For the present, with the exception of photoconductivity, it is appropriate to focus on optical constants as related to electronic structure.

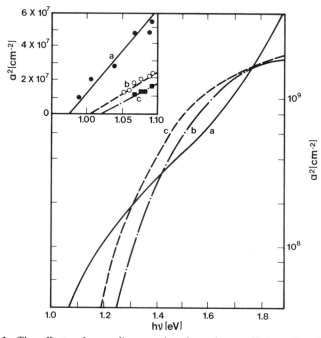

Fig. 4.1 The effects of annealing on the absorption coefficient of polycrystalline CuInSe$_2$ films. (a) As-deposited film (T_{sub} = 5.25 K; T_{se} = 540 K); (b) annealed in Ar (673 K, 30 min; (c) annealed in H$_2$Se/Ar (673 K, 30 min). Inset shows absorption edge with the lines representing a least-squares fit to the data (from Sun *et al.* [1]).

We shall further restrict ourselves to optical studies of polycrystalline semiconductor films which specifically address the effects of grain boundaries. As mentioned above, this eliminates most of the literature.

The salient feature of most optical data on polycrystalline semiconductor films is an additional absorption (compared to single crystals) for photon energies less than the band gap. This absorption may be obscured in indirect materials where the single-crystal absorption edge is not as steep as in direct gap materials. Furthermore, the amount of additional absorption generally depends on the grain size. Figure 4.1 illustrates this for the direct gap semiconductor $CuInSe_2$ [1].

This additional long-wavelength absorption might in principle increase the efficiency of a photovoltaic device, e.g., by absorbing a greater fraction of the solar flux in a solar cell. As discussed in other chapters of this volume, however, it is clear that grain boundaries in general have severe negative effects upon device performance. This is because the boundaries generally serve to limit both minority carrier lifetime and electron transport. The additional absorption thus is expected to have negligible effects on the efficiency.

In addition to a broadening of the absorption edge, the structure in the optical constants observed at higher photon energies will be less sharp in polycrystalline compared to crystalline semiconductors. These high-energy features may be more amenable to modeling than the band edge, although the experimental techniques required to obtain good data are more sophisticated.

4.2 EXPERIMENTAL TECHNIQUES FOR DETERMINING OPTICAL CONSTANTS

In this section we review frequently used procedures for the determination of optical constants of thin films. In attempting to measure small variations in optical constants due to polycrystallinity, one must be aware of the various artifacts and experimental errors which limit the precision of the measurement.

For completeness, we set down the relations between the various optical constants. For a derivation of these relations see Pankove [2].

The complex index of refraction n_c is defined as

$$n_c = n - ik \tag{4.1}$$

and is related to the velocity of propagation by

$$v = c/n_c \tag{4.2}$$

We will refer to n as the index of refraction and to k as the extinction coefficient. The absorption coefficient α is related to k by

$$\alpha = 4\pi k / \lambda \tag{4.3}$$

where λ is the wavelength of the light in a vacuum.

The dielectric constant can be introduced in two ways. One can define the dielectric constant ϵ to be real and describe any losses by a conductivity σ. Then

$$n^2 - k^2 = \epsilon \tag{4.4a}$$

$$nk = \sigma / \nu \tag{4.4b}$$

where ν is the frequency. Alternatively one can define ϵ to be complex:

$$\epsilon \equiv \epsilon_1 - i\epsilon_2 \tag{4.5}$$

Then

$$n^2 - k^2 = \epsilon_1 \tag{4.6a}$$

$$2nk = \epsilon_2 \tag{4.6b}$$

Clearly a knowledge of n and k determines ϵ_1 and ϵ_2, and vice versa. The quantity ϵ_2 is often more convenient to use in theoretical calculations, while experimental results often appear in terms of α. A plot of α versus photon energy has generally the same shape as a plot of ϵ_2, because n usually does not vary strongly with energy.

We now discuss two general methods for determining optical constants, leaving the problem of surface roughness and inhomogeneities until the end of this section.

1. Transmission and Reflectance at Normal Incidence

Consider monochromatic radiation incident normally upon an absorbing film on a transparent substrate. We assume initially that the substrate and film surfaces are perfectly smooth, that the film thickness is uniform, and that the optical constants of the film and substrate are homogeneous.

The transmission T and reflectance R of the system depend upon the optical constants n and k and the thickness d of the film, the wavelength λ of the light, and the index of refraction n_s of the substrate. The actual expressions are rather involved and will not be reproduced here, since they can be found in the literature [3, 4]. In general T and R both exhibit oscillations versus wavelength in wavelength regions where the film is transparent. A typical curve of transmission versus wavelength is shown in Fig. 4.2.

One is thus confronted with the task of solving the equations

$$T(n, k, d, \lambda) - T_{exp} = 0 \tag{4.7a}$$

$$R(n, k, d, \lambda) - R_{exp} = 0 \tag{4.7b}$$

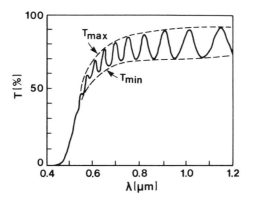

Fig. 4.2   Transmission spectrum at 300 K for an SnO_2 layer prepared by flash evaporation; $t = 2.03 \pm 0.08$ μm (from Manifacier *et al.* [10]).

where T_{exp} and R_{exp} are the experimentally determined values of T and R, respectively. Equations (4.7a) and (4.7b) are far too complex to permit solution in terms of n and k, so that with a few exceptions numerical procedures are required. During the last 10 years a variety of these procedures have evolved.

Typically one determines λ, n_s, and d independently. The thickness d may be determined using a variety of techniques. Bennett and Bennett [5] describe the various interferometric methods. Alternatively one can use a commericially available surface-profiling stylus. It is also possible to determine d from T in certain cases, as discussed below.

In regions of low absorption, one can determine n from the position of interference maxima or minima. For maxima one has

$$2n_1 d = m\lambda_1 \tag{4.8a}$$

$$2n_2 d = (m + 1)\lambda_2 \tag{4.8b}$$

where m is the order of the interference and λ_1 and λ_2 are the wavelengths of adjacent maxima. If one assumes that n is constant in this spectral region, then these equations determine n and m. One then proceeds to determine n at other wavelengths using Eq. (8a) and the appropriate m.

As k increases, Eq. (8a) eventually becomes invalid, and of course in regions of high k the interference structure disappears. Nevertheless one usually uses these values of n as an initial estimate. The value of n in the absorption region may be estimated by extrapolating from the low-absorption region. This is done either graphically or by fitting n to some reasonable function, e.g., $n = a + b\lambda^{-2}$ [6].

With these values of $n(\lambda)$, one then computer-fits the transmission function [Eq. (7a)] to find the appropriate k. Often the Newton–Raphson technique is used in making the fit [7].

Based on the quality of the film and the desired accuracy, the value of k found this way may be sufficient. If not, one must adjust the value of $n(\lambda)$ and repeat the process. This may be done by fitting n to the reflectance data and then using this improved n to find a new k. Alternatively one can simply search for the pair of n and k values which give the best fit to T [8]. Here one must have a reasonable initial estimate for n (say within 10%) to avoid spurious solutions.

Manifacier $et\ al.$ [9, 10] have developed a rather straightforward procedure for extracting n, k, and d from transmission measurements provided $k^2 \ll n^2$. In this method, one constructs envelope functions $T_{\max}$ and $T_{\min}$ (see Fig. 4.2) which are treated as continuous functions of λ. Then

$$n = \left[N + \left(N^2 - n_s^2\right)^{1/2} \right]^{1/2} \qquad (4.9)$$

with

$$N = \left(1 + n_s^2\right)/2 + 2n_s(T_{\max} - T_{\min})/T_{\max}T_{\min} \qquad (4.10)$$

The thickness is calculated from two maxima or two minima:

$$d = M\lambda_1\lambda_2/2\left[n(\lambda_1)\lambda_2 - n(\lambda_2)\lambda_1 \right] \qquad (4.11)$$

where M is the number of oscillations between the two extrema occurring at λ_1 and λ_2. With n and d determined, one may extrapolate n into the high-absorption region and determine k as before. This method is attractive because d is determined precisely in the region of the film where the transmission is measured. The authors claim a relative error in n for SnO_2 thin films which does not exceed 4–5% [9, 10]. For this method to be used most successfully, the films should be sufficiently thick (1–2 μm in the visible and near infrared) so that the interference peaks are closely spaced, thus defining $T_{\max}$ and $T_{\min}$ more precisely.

2. Ellipsometry

When the angle of incidence of the incoming light is not zero (in practice, when it is greater than about 10°), the reflected intensity is polarization-dependent. The reflectance R_p of light with the electric field parallel to the plane of incidence is different from the reflectance R_s with the electric field perpendicular to the plane of incidence. In addition there is generally a phase shift Δ between the two components.

Bennett and Bennett [5] discuss the various configurations used to extract optical constants of thin films using polarimetric measurements. In principle R_p and R_s or R_p/R_s may be measured at various angles, or the reflectance of unpolarized light may be measured at various angles. The precision obtained in a specific configuration depends upon the nature of the thin-film system under study. The reader is referred to the references in Bennett and Bennett [5] for details.

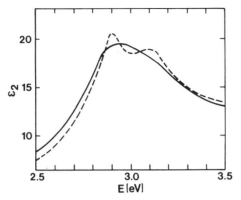

Fig. 4.3 E_1 and $E_1 + \Delta_1$ structures in ϵ_2 for relatively pure ($N_D = 1 \times 10^{16}$ cm^{-3}) and moderately doped ($N_D = 9 \times 10^{17}$ cm^{-3}) single crystals of GaAs at 300 K: ---, 1×10^{16} cm^{-3}; ———, 1×10^{17} cm^{-3} (from Aspnes [11]).

Ellipsometry consists of a measurement of the ratio of the reflected amplitudes:

$$\tan \psi \equiv r_p / r_s \tag{4.12}$$

together with the phase difference between r_p and r_s:

$$\Delta \equiv \delta_p - \delta_s \tag{4.13}$$

Aspnes [11] and Bashara and Azzam [12] have provided recent reviews of ellipsometry. This technique tended to be restricted to specialists until a few years ago because of (1) the computational efforts involved, and (2) the time and skill required to make precise measurements. The availability of small, inexpensive computers has spurred development of automated instrumentation, some of which is now commercially available. This permits one to consider the use of ellipsometry as a spectroscopic tool, rather than for measuring the thickness of an overlayer using a fixed wavelength.

Ellipsometry is generally capable of considerably more sensitivity and precision than intensity measurements for the determination of optical constants. For example, one can detect the presence of a fraction of a monolayer of an absorbed substance on a substrate [13]. Alternatively one can achieve high precision in the determination of optical properties of bulk material. An example of this is shown in Fig. 4.3.

The analysis and procedures involved in determining n and k from ellipsometric data are complicated and not within the scope of the present review. The reader is referred to the reviews cited above for details.

3. The Effects of Surface Roughness and Inhomogeneities

The preceding discussion assumed a perfectly smooth film on a perfectly smooth substrate. In fact, for the majority of the literature on optical properties of thin films, such an assumption is implicit. Transmission and reflectance may be measured in a fairly straightforward manner to about 1% accuracy using commercial instruments. The accuracy may be extended to about 0.1% if special precautions (e.g., linearity of response, reduction of stray light) are taken.

The uncertainties in n and k are complicated functions of the wavelength dependence and the thickness of the film. By examining the dominant terms in the expressions for R and T, however, one sees that 1% uncertainties in R and T produce order-of-magnitude uncertainties of 1% in n and k (for a 1 μm thick film of a typical semiconductor near the main absorption edge). One may *not* conclude from this, however, that the optical constants have been measured to this accuracy. The difficulty is that this treatment neglects surface roughness.

Several authors treat the effects of surface roughness on the reflectance and transmission (coherent and incoherent) of bulk surfaces and thin films [4, 5, 14].

For normal incidence the ratio of reflectance from a rough surface relative to a smooth surface of the same material is given by

$$R/R_0 = \exp\left[-(4\pi\sigma/\lambda)^2\right] + \left\{1 - \exp\left[-(4\pi\sigma/\lambda)^2\right]\right\}$$
$$\times \left\{1 - \exp\left[-2(\pi\sigma\alpha/m\lambda)^2\right]\right\} \qquad (4.14)$$

where σ is the rms height of the surface irregularities, λ the wavelength, m the rms slope of the irregularities, and α the half acceptance angle of the instrument. This expression is valid for $\sigma/\lambda \ll 1$. The first term is due to coherently reflected light, and the second term is due to that part of the incoherently reflected light recorded by the instrument. Bennett and Bennett [5] have discussed the effect of surface roughness on the accuracy of reflectance measurements. They show that, to obtain 0.1% accuracy in a reflectance measurement, σ must be on the order of 10 Å for visible wavelengths. In general special polishing techniques are necessary to achieve such smoothness. Alternatively one may attempt to measure the roughness and correct for it in the reflectance determination.

Szczyrbowski and Czapla [4] have studied the effects of surface roughness on R and T for a thin film on a smooth transparent substrate. Their analysis includes the effects of multiple reflections. They then apply these results to a determination of n and k for polycrystalline InAs films. As an example, Fig. 4.4 shows the absorption coefficient for a 3.5-μm-thick film, first assuming $\sigma = 0$ and then using a best-fit value, $\sigma = 0.10$ μm. Clearly the effect of surface roughness is significant at low absorption coefficients and must be accounted for if one hopes to understand such small effects as those due to grain boundaries.

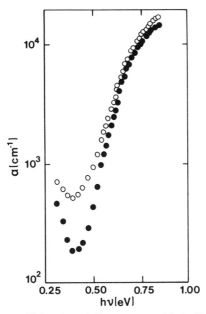

Fig. 4.4 Absorption coefficient for a 3.51-μm sputtered InAs film showing the effects of neglecting surface roughness. O, $\sigma = 0$; ●, $\sigma = 0.10$ μm (from Szczyrbowski and Czapla [4]).

In analogy with surface roughness, spatial fluctuations of the optical constants in the bulk of a material can also lead to anomalous R and T values. The presence of grain boundaries and other structural imperfections may be expected to lead to such fluctuations. This is a more complicated problem then surface roughness, however, and it has not been extensively studied theoretically. Filinski [14] has presented general expressions for R and T due to a bulk sample including both surface roughness and bulk spatial fluctuations. They are

$$R = R_s \exp\left[-(4\pi\sigma n_0/\lambda_0)^2 - (\Delta\theta_{IIR})^2 \right] \tag{4.15}$$

$$I = I_0(1 - R_s)^2 \exp\left\{ -\Gamma d - \left[2\pi(N - n_0)\sigma/\lambda_0 \right]^2 - (2\pi\Delta L/\lambda_0)^2 \right\} \tag{4.16}$$

where R_s is the reflection coefficient for a smooth surface:

$$R_s = \left[(N - 1)^2 + K^2 \right] / \left[(N + 1)^2 + K^2 \right] \tag{4.17}$$

$$N = n_M n_{eff} \tag{4.18}$$

$$\Gamma = \gamma_0\left(2n_{eff}^2 - 1\right) / n_{eff} + \langle\Delta\epsilon^2\rangle n_M^4 \omega^4 a^3 / 12\pi n_{eff} c^4 \tag{4.19}$$

$$n_{eff} = \left(1 + \mu\langle\Delta\epsilon^2\rangle n_M^2 \omega^2 a^2 / 6\pi c^2\right)^{1/2} \tag{4.20}$$

$$\Delta L = \sqrt{2}\, \sigma(n - n_0) \tag{4.21}$$

In these expressions, $\Delta\theta_{\mathrm{IIR}}$ is the rms value of the fluctuation in the phase due to the fluctuating dielectric constant, σ the rms surface roughness, n_0 the index of the medium of the incident beam, n_{M} the mean index of the sample, γ_0 the mean absorption coefficient, $\langle\Delta\epsilon^2\rangle$ the mean square fluctuation of the dielectric constant, a the fluctuation correlation length, and μ a number depending upon the form of the correlation function. These expressions are valid for a small correlation length, $n_{\mathrm{M}}\omega a/c \ll 1$, and weak absorption, $n_{\mathrm{M}}\omega/c \gg \gamma_0$.

Filinski [14] has studied theoretically the effects of such inhomogeneities on the reflectance of classical oscillators. He finds sharp structure in the reflectance curves, of the kind often attributed to intrinsic properties of a solid (e.g., polarition effects). To the present author's knowledge, an attempt to model a solid containing grain boundaries using this treatment has not been undertaken.

4.3 MODELS OF THE OPTICAL PROPERTIES OF A POLYCRYSTALLINE SEMICONDUCTOR

We now turn specifically to the optical properties of polycrystalline semiconductors, focusing on the absorption edge. It should be emphasized at the outset that it may not be possible to fabricate a real polycrystalline semiconductor composed solely of high-quality bulk material separated by grain boundaries. (This point is discussed further in Chapters 3 and 9.) The material is apt to contain unwanted and uncontrolled impurities, stoichiometry deviations, point defects, etc., in addition to grain boundaries. In general, optical properties will be less sensitive than electrical properties to these effects, but the same statement will in general be true for the grain boundaries. Thus agreement of models with experiment should be regarded as tentative at best.

A new generation of facilities for evaporation is rapidly coming into use, based upon molecular beam epitaxy systems. Here one grows the semiconductor film in an ultrahigh vacuum environment (10^{-10} torr base pressure) and characterizes many properties *in situ*, using, for example, electron diffraction, Auger, and SIMS analysis. Polycrystalline films grown in such systems may be expected to be significantly more free of unwanted impurities and possibly some structural defects. Such films may permit more realistic modeling of optical and electrical properties.

1. Absorption Coefficient in Single-Crystal Semiconductors

The dominant feature of the energy dependence of the absorption coefficient $\alpha(h\nu)$, and the one to which we address ourselves, is the onset of absorption near the region of interband transitions from valence to conduction bands. Only the basic results are given here. The reader is referred to several useful texts for details, including those by Pankove [2], Greenaway and Harbeke [15], and Bassani and Pastori Parravicini [16].

We summarize for convenience the energy dependence of α near the band edge for band-to-band and exciton transitions:

(1) *Allowed direct transitions*, neglecting exciton effects [2]:

$$\alpha(h\nu) \sim (h\nu - E_g)^{1/2} \tag{4.22}$$

where E_g is the energy gap at $k = 0$.

(2) *Forbidden direct transitions*, neglecting exciton effects [2]:

$$\alpha(h\nu) \sim (h\nu - E_g)^{3/2}/h\nu \tag{4.23}$$

(3) *Indirect transitions*, neglecting exciton effects [2]:

$$\alpha(h\nu) = \alpha_e(h\nu) + \alpha_a(h\nu) \tag{4.24}$$

where

$$\alpha_e(h\nu) = A\left(h\nu - E_g - E_p\right)^2 / \left[1 - \exp(-E_p/kT)\right] \tag{4.25}$$

for $h\nu > E_g - E_p$, α_e corresponds to the emission of a phonon of energy E_p in order to conserve momentum, and

$$\alpha_a(h\nu) = A\left(h\nu - E_g + E_p\right)^2 / \left[\exp(E_p/kT) - 1\right] \tag{4.26}$$

corresponding to the absorption of a phonon.

(4) *Inclusion of exciton effects*: In sufficiently pure semiconductors, the intrinsic absorption is dominated by exciton formation, giving rise to sharp line structure in direct gap materials at low temperature and to a more complicated expression for $\alpha(h\nu)$ at all temperatures. The explicit dependence of α on $h\nu$ does not concern us here. The interested reader is referred to Chapter 6 of Bassani and Pastori Parravicini [6]. Exciton effects are important, however, in understanding the role of grain boundaries on $\alpha(h\nu)$. We return to this point later.

In addition to band-to-band absorption, a single crystal may exhibit impurity effects in its absorption spectrum. These effects include acceptor–conduction band, valence band–donor, and possibly acceptor–donor transitions, all on the low-energy side of the absorption edge. For low to moderate doping ($< 10^{18}$ cm^{-3}) these effects will be small compared to band-to-band absorption. For heavy doping, complications arise because of the onset of degeneracy, impurity banding, and potential fluctuations which generally affect the shape of the main absorption edge. (See [2] for a discussion of various impurity effects.) Since many polycrystalline semiconductors are in this class, it is seen that such complications make it difficult to assess the effects of grain boundaries on the absorption edge in a real crystal.

An indirect gap semiconductor (e.g., Si or Ge) has a rather broad absorption edge [Eq. (4.24)] relative to that of a direct gap semiconductor (e.g., GaAs, CdS) [Eq. (4.22)]. Thus small modifications of the absorption edge due to grain boundaries will be most apparent for a direct gap semiconductor with a sharp intrinsic edge. To be sure, there is considerable literature showing variations in the shape of the absorption edge in indirect semiconductors as a function of degree of crystallinity [17]. It is not clear from such studies, however, that grain boundaries are solely responsible for the shape. The same criticism applies of course to direct gap semiconductors, but here the sharpness of the intrinsic absorption edge permits one to study small deviations more easily.

To the writer's knowledge, there have been only four published attempts to model the contribution of grain boundaries to optical absorption. We proceed to discuss these models.

2. Polycrystalline Semiconductors as a Mixture of Amorphous and Single-Crystal Materials

Szczyrbowski and Czapla [4] have attempted to describe the optical absorption of sputtered InAs films by a model in which the film is composed of microcrystallites embedded in an amorphous matrix. No evidence is presented that this is an appropriate structural model. One might, however, divide the sample into a volume associated with grain boundaries and a volume associated with bulk regions. For example, if one has a 1000-Å crystallite and the width of the grain boundary is assumed to be 10 Å, then the grain boundary volume is a few percent of the total volume, which is not inconsistent with their results.

The assumption that the grain boundary region is amorphous is tenuous. On the other hand, all that is required for optical absorption in this model is that the transition be nondirect; i.e., k conservation is not important [18]. The grain boundary could be sufficiently disordered such that the wave vector is not a good quantum number, even though it might not be, strictly speaking, in an amorphous state. The situation is further complicated by the realization that the grain boundary is a sink for various impurities. Thus, for example, significant amounts of oxygen in the grain boundary would appreciably modify the effective energy gap.

Szczyrbowski and Czapla fit their absorption data to the expression

$$\alpha = \beta\alpha_{cr} + (1 - \beta)\alpha_{am} \qquad (4.27)$$

where β is the volume fraction of the crystalline phase, α_{cr} is essentially Eq. (4.22) modified to account for degeneracy and nonparabolicity, and α_{am} is given by the nondirect transition expression of Tauc [18]:

$$\alpha_{am}h\nu - \left(h\nu - E_g^{am}\right)^2 \qquad (4.28)$$

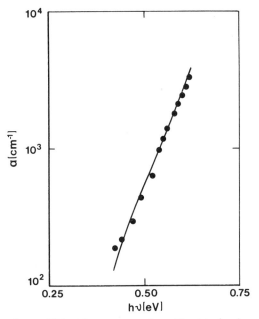

Fig. 4.5 Absorption coefficient for same sample as Fig. 4.4, showing a fit to the model consisting of a single crystal–amorphous mixture. See text for details (from Szczyrbowski and Czapla [4]).

again modified to account for nonparabolicity. Figure 4.5 shows a best fit for one of their films. They find $\beta = 0.95$, $E_g^{cr} = 0.35$ eV, and $E_g^{am} = 0.33$ eV. Their analysis also yields an electron concentration (7.5×10^{18} cm^{-3}) in reasonable agreement with that obtained from the plasma resonance (6.0×10^{18} cm^{-3}).

Bagley *et al.* [19] have also applied the amorphous–single crystal mixture model to the optical properties of low-pressure chemical vapor-deposited silicon. They used the effective medium approximation [20] to account for ϵ_1 and ϵ_2 at energies (3.0–5.8 eV) well above the fundamental absorption edge. In contrast to the work on InAs, however, this system is known from structural studies to contain crystalline and amorphous regions. The authors found, in undoped material, that the best fit was given by a mixture of 64% amorphous Si, 20% crystalline Si, and 16% voids.

3. The Dow–Redfield Model

In a series of papers [21–24], Dow and Redfield have shown that the excess absorption (usually an exponential tail) occurring on the low-energy side of the absorption edge in semiconductors is due to a Franz–Keldysh effect arising from electric fields present in the material. The electric fields (typically on the order of 10^5 V/cm) arise from a variety of sources, such as charged impurities, external surfaces, and grain boundaries.

The Franz–Keldysh effect is basically photon-assisted tunneling. In the presence of an electric field, the electron has a nonzero probability of being in the forbidden gap. Thus a photon with energy less than E_g can cause a transition to the conduction band. The absorption coefficient α as a function of field F is given by [25]

$$\alpha = I(F)F^{1/3} \qquad (4.29)$$

where

$$I = K\nu^{-1} \int_y^\infty \mathrm{Ai}^2(z)\,dz \qquad (4.30)$$

$$y = \gamma^{1/3} F^{-2/3}(E_g - h\nu) \qquad (4.31)$$

Ai is the Airy function, while γ and K are constants of the material.

The above relations do not account for exciton effects. Dow and Redfield [22] have shown that the Coulomb attraction between the electron and hole causes a significant reduction in the tunneling barrier. Numerical solutions show that the sub-band gap absorption is enhanced by several orders of magnitude by the inclusion of exciton effects. In spite of this, the shape of the absorption curve is approximately the same for $E_g - h\nu$ greater than about one or two exciton Rydbergs (0.035 eV in CdS, 0.055 eV in ZnS). This is seen in Fig. 4.2 from Dow and Redfield [23]. Since the models to be discussed in the next section are normalized to the absorption at the energy gap, the neglect of exciton effects is not critical.

Since the electric fields responsible for this enhanced absorption are usually nonuniform and random, the absorption coefficient for a given energy must be averaged over the field. One may account for this in two ways. Dow and Redfield [23] used a uniform microfield approximation:

$$\langle \alpha(h\nu) \rangle = \int_0^\infty P(F)\alpha(h\nu, F)\,dF \qquad (4.32)$$

where $P(F)$ is the probability that there is a field F in the material. Bujatti and Marcelja [26] calculated the total absorption in the sample

$$A = \frac{1}{S} \int_V \alpha(\mathbf{r}, F)\,dV \qquad (4.33)$$

where S is the area of the illuminated surface.

4. Application of the Dow–Redfield Model to Polycrystalline Semiconductors

Bujatti and Marcelja [26] applied the Dow–Redfield model to the absorption edge in polycrystalline CdS films. Their films were evaporated, some being exposed to air during depositon. Presumably the presence of O increased the charge on the grain boundaries. The samples were divided into two groups: (1) composed of 0.1 to 0.5-μm-sized cystallites in addition to smaller crystallites, and (2) composed only of the smaller crystallites. Electrical measurements showed the presence of potential barriers on the order of 0.3 eV in type-1 films.

In approximating the effect of the electric field, Bujatti and Marcelja assumed spherical crystallites with some distribution $P(R)$ of the radius of the spheres. Equation (4.33) then becomes [26]

$$A = \frac{4\pi N}{S} \int_0^\infty P(R)\,dR \int_{r_m}^R \alpha(\mathbf{r}, F) r^2 \, dr \qquad (4.34)$$

where N is the number of crystallites in the volume V and r_m is the minimum radius, measured from the center of the crystallite, at which absorption can occur for a given field and photon energy. Two limiting cases are then considered to make the problem tractable.

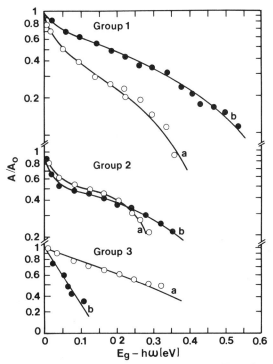

Fig. 4.6 Absorption coefficient for various CdS polycrystalline films. The absorption coefficient is normalized to 1 at the energy gap. The solid lines are fits to the Dow–Redfield model. ●, Films grown in high vacuum; ○, films exposed to air during deposition. See text for details (from Bujatti and Marcelja [24]).

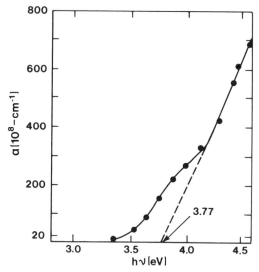

Fig. 4.7 The square of the absorption coefficient in sputtered ZnS polycrystalline films (from Bugnet [25]).

(1) The number of acceptorlike surface states is so large (and the crystallite is so small) that the donors in the crystallite are fully ionized, giving a constant charge density. This gives a radially increasing field within the crystallite.

(2) The crystallites are sufficiently large that only donors near the surface are ionized. In this case the field decays exponentially from the surface. The field-dependent absorption coefficient [Eq. (4.29)] is then calculated and suitably averaged using the assumptions stated above. A comparison of the model with experiment is shown in Fig. 4.6. Group 1 is composed of films with an unusually high percentage of large crystallites (10^{-5} cm) almost covering the entire surface. Group 2 is composed of films with only a few such large crystallites, while group 3 is composed of films with only very small crystallites (less than 500 Å).

The films in group 1 are amenable to analysis. Bujatti and Marcelja [26] used the exponential field approximation (Debye length 10^{-5}–10^{-6} cm) and found a surface charge density of 10^{12}–10^{13} cm^{-2} and potential barriers of 0.3–0.4 eV. These results are consistent with results of electrical measurements. The films in groups 2 and 3 were fit using a sum of exponential and linear field terms to account for the smaller crystallites. The parameters extracted from these fits do not yield quantities which are particularly meaningful.

Bugnet [27] made a similar analysis on reactively sputtered polycrystalline ZnS films. His results are shown in Fig. 4.7. He found that the excess absorption at the band gap (3.7 eV) could be attributed to the Franz–Keldysh effect with a surface potential of 1.36 V at the grain boundary. This potential seems rather high, but it is not inconsistent with the value of 1.2 V found by Swank [28] from contact potential measurements. At longer wavelengths, however, the excess absorption cannot be accounted for by this effect. Furthermore, the long-wavelength absorption is very sensitive to preparation conditions compared to absorption near the band edge. Bugnet suggested that this additional absorption was due to other defects and impurities [27].

4.4 CONCLUSIONS

For a number of reasons, it is apparent that progress in understanding optical properties of polycrystalline semiconductors has been rather meager. The effects tend to be small and, because of a traditional emphasis on single-crystal devices, they have been viewed to be of little practical or fundamental interest. Furthermore, effects due strictly to grain boundaries are often inseparable from those due to other defects and impurities. Finally, most optical measurements have not adequately accounted for surface and substrate roughness effects. The neglect of these effects has added additional uncertainty to the interpretation of results.

The development of spectroscopic ellipsometry promises more precise determination of optical constants, particularly at photon energies well above the fundamental edge. It may be possible to model the effects of grain boundaries more adequately at high energies than at the absorption edge.

The models described in the previous section represent interesting attempts at understanding this problem. The Dow–Redfield model applied to charges at grain boundaries is certainly a realistic approach to understanding absorption near the band edge. The materials studied thus far, however, have been insufficiently characterized to permit a clear-cut interpretation of the results.

Clearly studies are required on polycrystalline materials of higher quality. There is a limitation on this, however, because processes responsible for polycrystalline structure may also lead intrinsically to other structural defects. In addition the grain boundaries may serve as a sink for various unwanted impurities, although some passivation may be possible. The proliferation of molecular beam epitaxy facilities during the next few years, coupled with a complement of analytical tools for characterizing such materials, should enable investigators to fabricate polycrystalline films of sufficient quality in order to better understand their optical properties.

ACKNOWLEDGMENTS

The author profited from useful discussions with Drs. Brian Bagley, Bell Laboratories, Edward Fagen, University of Delaware, and David Redfield, RCA Laboratories.

REFERENCES

1. L. Y. Sun, L. L. Kazmerski, A. H. Clark, P. J. Ireland, and D. W. Morton, *J. Vac. Sci. Technol.* **15**, 265 (1978).
2. J. Pankove, "Optical Processes in Semiconductors." Prentice-Hall, Englewood Cliffs, New Jersey, 1971.
3. O. S. Heavens, "Optical Properties of Thin Solid Films." Butterworths, London, 1955.
4. J. Szczyrbowski and A. Czapla, *Thin Solid Films* **46**, 127 (1977).
5. H. E. Bennett and J. M. Bennett, "Physics of Thin Films" (G. Hass and R. E. Thun, eds.) Vol. 4, pp. 1–96. Academic Press, New York, 1967.
6. T. S. Moss, "Optical Properties of Semiconductors." Butterworths, London, 1959.
7. I. S. Berezin and N. P. Zhidkov, "Computing Methods," Vol. II. Addison-Wesley, Reading, Massachusetts, 1965.
8. J. Wales, G. J. Lovitt, and R. A. Hill, *Thin Solid Films* **1**, 137 (1967).
9. J. C. Manifacier, J. Gasiot, and J. P. Fillard, *J. Phys. E* **9**, 1002 (1976).
10. J. C. Manifacier, M. DeMurcia, J. P. Fillard, and L. Vicario, *Thin Solid Films* **41**, 127 (1977).
11. D. E. Aspnes, "Optical Properties of Solids—New Developments" (B. O. Seraphin, ed.), pp. 799–846. North-Holland Publ., Amsterdam, 1976.
12. N. M. Bashara and S. C. Azzam (eds.), "Ellipsometry." North-Holland Publ., Amsterdam, 1976.
13. F. Meyer, "Ellipsometry" (N. M. Bashara and S. C. Azzam, eds.), pp. 37–48. North-Holland Publ., Amsterdam, 1976.
14. I. Filinski, *Phys. Status. Solidi (b)* **49**, 577 (1972).
15. D. L. Greenaway and G. Harbeke, "Optical Properties and Band Structure of Semiconductors." Pergamon, Oxford, 1968.
16. F. Bassani and G. Pastori Parravicini, "Electronic States and Optical Transitions in Solids." Pergamon, Oxford, 1975.
17. A. Glass, *Can. J. Phys.* **43**, 1068 (1965).
18. J. Tauc, "Optical Properties of Solids" (F. Abeles, ed.), pp. 277–313. North-Holland Publ., Amsterdam, 1972.
19. B. G. Bagley, D. E. Aspnes, and C. J. Mogab, *Bull. Am. Phys. Soc.* **24**, 363 (1979).
20. R. Landauer, in "Electrical Transport and Optical Properties of Inhomogeneous Media" (J. C. Garland and D. B. Turner, eds.), pp. 2–45. American Institute of Physics, New York, 1978.
21. D. Redfield, *Phys. Rev.* **130**, 916 (1963).
22. D. Redfield and M. A. Afromowitz, *Appl. Phys. Lett.* **11**, 138 (1967).
23. J. D. Dow and D. Redfield, *Phys. Rev. B* **1**, 3358 (1970).
24. J. D. Dow and D. Redfield, *Phys. Rev. B* **5**, 594 (1972).
25. J. Callaway, *Phys. Rev.* **134**, A998 (1964).
26. M. Bujatti and F. Marcelja, *Thin Solid Films* **11**, 249 (1972).
27. P. Bugnet, *Rev. Phys. Appl. Fr.* **9**, 447 (1974).
28. R. K. Swank, *Phys. Rev.* **153**, 844 (1967).

5 The Electronic Structure of Grain Boundaries in Polycrystalline Semiconductor Thin Films

LEWIS M. FRAAS   and   KENNETH ZANIO

Chevron Research Company
Richmond, California

Hughes Research Laboratories
Malibu, California

5.1 INTRODUCTION

The revolution in solid-state electronics began with a theory of a perfect periodic lattice with donor or acceptor impurities. This theory allowed the development of bulk-effect devices such as junction diodes and bipolar transistors. Next, extension of the theory of the solid state to surfaces paralleled the development of surface-effect devices such as field-effect transistors (FETs) and charge-coupled devices (CCDs). In these two areas, semiconductor process development led to successful commercial devices because the solid-state theory provided predictability.

Polycrystalline semiconductor thin-film development, on the other hand, has proceeded by correlating process variables with device variables

POLYCRYSTALLINE AND AMORPHOUS
THIN FILMS AND DEVICES

without the aid of a theory allowing predictability. Success has been limited at best. In this paper, we take the point of view that solid-state theory must be extended from the bulk periodic lattice and the external interfaces to describe the one- and two-dimensional defect arrays within polycrystalline thin films (i.e., dislocations, stacking faults, and grain boundaries). Since these defect arrays can, and generally do, have associated electronic banding states within the bulk material energy gap, it is desirable to characterize these states (i.e., describe their origins, impurity interactions, energy-level positions, state densities, and effects on device performance). However, very little work has been done along these lines. But work can begin in the sense that the experimental and theoretical techniques are available. An integrated point of view is required.

In this paper, we attempt to provide the required integration by first describing the origins of the one- and two-dimensional defect bands (Section 5.2). Then, we relate the defect bands to thin-film device performance (Section 5.3) and describe means by which these defect bands can be experimentally measured (Section 5.4). Finally, we suggest ways that the defect bands might be modified by technology to lead to grain boundary and dislocation passivation (Section 5.5).

5.2 INTERNAL BOUNDARIES, ELECTRONIC STRUCTURE, AND ELECTRONIC TRANSPORT

A picture of the electronic structure of dislocations and free surfaces has developed from experimentation over the past few years. This section begins with a review of this work. Then the theory of dislocation states and surface states is generalized to describe the electronic structure of grain boundaries. This is followed by a description of electronic conduction perpendicular and parallel to grain boundaries. A later section discusses electronic transport across a $p-n$ junction in the region where a grain boundary intersects this junction.

1. Electronic Structure of a Dislocation

A dislocation is a one-dimensional periodic defect array. Dislocations can be seen experimentally in many ways. Figure 5.1 presents two microscopic photographs of dislocations. Figure 5.1a shows a cathodoluminescence [1] image of dislocations in GaAs. The black lines represent the dislocations. The one-dimensional periodicity is evident, as the dislocations run as far as 50 μm. Figure 5.1b is a transmission electron microscope photograph of the core region [2] of a 60° dislocation in germanium as viewed along the dislocation axis. The fringe separation in this photograph

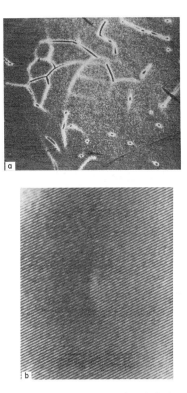

Fig. 5.1 (a) Cathodoluminescence image of Te-doped GaAs. A number of dislocations are visible as black lines and dots where the recombination of electron–hole pairs induced by electron bombardment occurs locally without radiation. Magnification 145 × (from Schiller and Boulou [1]). (b) ($\bar{1}$11) lattice plane image of 60° dislocation in (110) foil from twisted germanium crystal. Fringe spacing 3.266 Å (from Phillips and Wagner [2]).

represents 3.27 Å. In general, disruption of the single-crystal periodic lattice near the dislocation core will lead to localized dislocation states at the core, and the one-dimensional periodicity along the dislocation axis will lead to banding of these states [3].

The origins of the electronic states in the dislocation core region will now be addressed. Figure 5.2 shows three atomic models for the dislocation core region in the diamond lattice [4]. In Fig. 5.2a and b, the 60° and (100) edge dislocations possess both dangling bond and distorted bond regions. Figure 5.2c shows a form of the (100) edge dislocation in which the dangling bond region has been healed. From these atomic arrangements, we can identify two origins of electronic state (i.e., the dangling bond sites and the distorted bond regions). A third origin of dislocation electronic states will result from disruption in the one-dimensional periodic

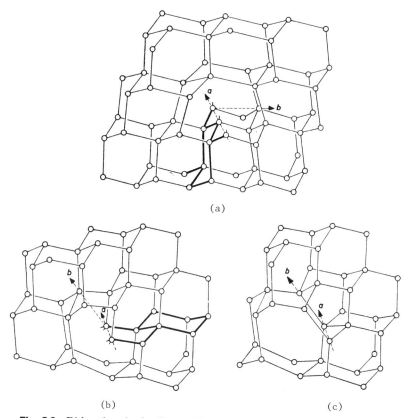

Fig. 5.2 Dislocations in the diamond lattice. (a) dislocation, 60°; (b), (c) two forms of the edge dislocation with glide plane (100). **a**, Axis; **b**, Burger's vector; heavy lines denote extra half planes (from Hornstra [4]).

array along the dislocation axis. Sources of such a disruption might be kinks, tangles, jogs, or precipitates.

Next we rank these state origins in order of importance. The experimental data for germanium show that the distorted bands are of first-order importance in describing the electronic states associated with germanium dislocations. Although it is clear that the disruption in the one-dimensional periodicity can be treated in second order, it is not clear a priori that the dangling bond states can be treated in second order. However, the preponderance of experimental data shows little evidence of dominant dangling bond states in the dislocation cores of Ge and Si [3]. Electron spin resonance (ESR) experiments, for example, done on crystals before and after deformation show that the number of active resonance centers is about one order of magnitude smaller than the expected number of free

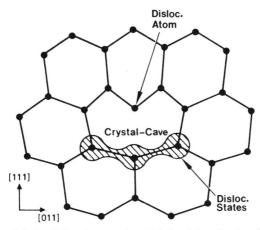

Fig. 5.3 Model of a dislocation core in which the dislocation band states are attributed to the dilated bonds shown (from W. Barth *et al.* [7]).

bonds attributed to the dislocations as revealed by etch pit counts. We take the point of view, then, that the dangling bond sites in dislocation cores are chemically very active sites and, therefore, that they are easily healed by gettering of reactive impurities (e.g., hydrogen or oxygen) from the crystal bulk. If, for example, a hydrogen atom were to attach to a dangling bond, the dangling bond state would be stabilized and removed from the band gap region much as occurs in the amorphous hydrogenated silicon system. This hypothesis is consistent with the experimental ESR observation that the concentration of unpaired electrons decreased with an increase in the crystal deformation temperature or with subsequent annealing [5].

Therefore, we will focus on the distorted bonds as the main mechanism generating defect states in the dislocation core. Figure 5.3 shows the hypothesized core states associated with the distorted (dilated) bond region of a dislocation. The thesis is that dilated bond lengths pull states out of the conduction and valence bands. Figure 5.4 shows how this might come about. The figure shows the theoretical boundaries of the energy bands of the diamond lattice with the lattice parameter as the variables [6]. For infinite atomic spacing, the group IV atoms have the ns^2np^2 configuration. As one imagines moving the atoms closer, the s^2 and p^2 bands cross and the tetrahedral sp^3 bonding orbitals take over. Then, as the lattice contracts, the valence and conduction bonding and antibonding states move further apart. Applying this model to bond dilation or contraction at the dislocation core leads to the conclusion that dilated bonds will pull states out of the valence and conduction bands. Hence, the arrangement in Fig. 5.3.

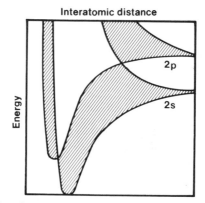

Fig. 5.4 Boundaries of energy bands of diamond as functions of interatomic distance (from Slater [6]).

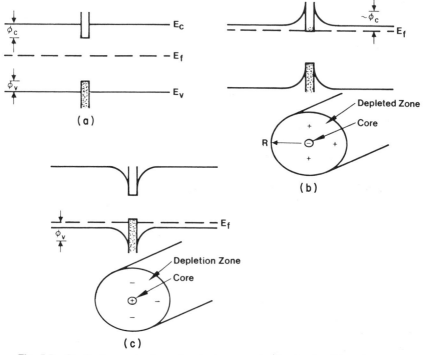

Fig. 5.5 Conduction and valence band edges associated with the dislocation core and adjacent single-crystal regions. (a) When the single-crystal region is undoped, the Fermi level is at the band gap center in both the core and single-crystal regions. (b) When the single-crystal region is doped *n* type, donor-state electrons can fall into the conduction band well at the dislocation core, resulting in a positively charged depletion region around the negatively charged core. (c) Similar to (b) but with opposite polarity.

We have discussed the origins of the dislocation band states. Now we note that the interaction of these states with the adjacent single-crystal region will lead to depletion regions adjacent to the dislocation core. This situation is shown in Fig. 5.5. The radius of the depletion region R and the dislocation level (trap level) filling can be calculated from Gauss' law given the dopant concentration, N_d or N_a, and the defect-level positions relative to the conduction and valence band edges, ϕ_c and ϕ_v. The resultant equations are

$$N_+ = \pi R^2 N_d \tag{5.1}$$

$$\phi_{c,\,v} = \frac{qn_+}{4\pi\epsilon_0}\left[\ln\left(\frac{n_+}{\pi N_{d,\,a}a^2}\right) - \left(1 - \frac{n_+}{\pi N_{d,\,a}a^2}\right)\right] \tag{5.2}$$

Let us now turn to a description of the experimental data supporting this picture of the electronic bands associated with the dilated bonds in the dislocation core region of germanium. The optical absorption [7], electroluminescence [8], Hall-effect measurement [9], ESR [5], and C–V data [10] are all consistent with a dislocation band model. All these data were obtained in controlled experiments where deformed crystals were compared with undeformed crystals. The signals studied correlated with the amount of deformation. We review the optical absorption [7] and electroluminescence [8] data. If we define the energy at the edge of the dislocation conduction band by E_{dc} and the energy at the edge of the dislocation valence band by E_{dv}, then in absorption experiments we would expect to see the $E_{dc} \rightarrow E_c$ transition (0.12 eV) only in *n*-type material and the $E_{dv} \rightarrow E_v$ transition (0.1 eV) only in *p*-type material. This is the observed result (Fig. 5.6). One would also expect to see an $E_{dv} \rightarrow E_{dc}$ transition (0.5 eV) that shifted slightly with dopant type or concentration. This is also observed (Fig. 5.7). One would expect a predictable variation [via Eqs. (5.1) and (5.2)] in absorption strength for, say, the $E_{dc} \rightarrow E_c$ transition as the dislocation density or carrier concentration is varied. Experimental

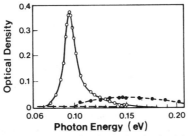

Fig. 5.6 Influence of doping type on the absorption in the 0.1-eV region of Ge at $T = 77$ K having an etch pit density of 3×10^7 cm^{-2} and sample thickness 4.1 mm. ●, $n = 1.5 \times 10^{16}$ cm^{-3}; dashed line, $p = 2.5 \times 10^{16}$ cm^{-3}; ○, intrinsic (from Barth *et al.* [7]).

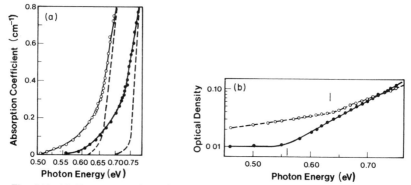

Fig. 5.7 (a) Temperature dependence of the absorption in the 0.6-eV region of n-Ge. The absorption of an undeformed sample is given by the dashed lines. $\bullet$, $T = 300$ K; $\bigcirc$, $T = 77$ K; $N_D = 2 \times 10^{16}$ cm^{-3}; $N_V = 1 \times 10^8$ cm^{-2} (from Barth et al. [7]). (b) Logarithmic plot of the absorption of n- and p-doped germanium samples (thickness, 4.1 mm). $T = 77$ K. The different threshold energies are indicated. $\bullet$, $p = 2.5 \times 10^{16}$ cm^{-3}, 3.7% deformed; $\bigcirc$, $n = 1.5 \times 10^{16}$ cm^{-3}, 3.2% deformed (from Barth et al. [7]).

data (Fig. 5.8) are consistent with such predictions. Finally, if the dislocation bands result from dilated bonds parallel to Burger's vector, some polarization dependence might be expected in the absorption spectra. The $E_{dc} \rightarrow E_c$ transition does, in fact, exhibit the expected pronounced dichroism. Finally, given the above model, one might expect to see electroluminescence in p–n junctions associated with the $E_{dc} \rightarrow E_{dv}$ transition (0.5 eV). And, as shown in Fig. 5.9, this is also observed.

The data described above are quite convincing for Ge. We see no reason why this picture cannot be generalized qualitatively to other semiconductors. Some indication that this might be valid is provided by the

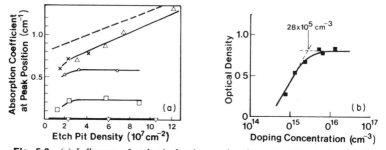

Fig. 5.8 (a) Influence of etch pit density on the absorption line in the 0.1-eV region of n-Ge at $T = 77$ K. The n-doping concentration is kept constant. The solid lines mark samples having the same doping concentration. $\triangle$, 7×10^{15} cm^{-3}; $\times$, 1.5×10^{16} cm^{-3}; $\bigcirc$, 1.5×10^{15} cm^{-3}; $\square$, 6×10^{14} cm^{-3}; ∇, intrinsic (from Barth et al. [7]). (b) Influence of doping concentration on the absorption line in the 0.1-eV region if the etch pit density is kept constant (etch pit density, 4.5×10^7 cm^{-2}) (from Barth et al. [7]).

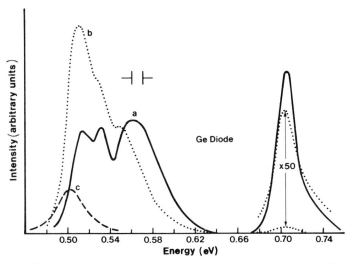

Fig. 5.9 Dependence of the emission spectra on the dislocation density N_v (injection current density j being constant). (a) $N_v = 4 \times 10^4$ cm^{-2}; (b) $N_v = 1 \times 10^6$ cm^{-2}; (c) $N_v = 1 \times 10^8$ cm^{-2}; $N_a = 3 \times 10^{16}$ cm^{-3}; $j = 4.1$ A-cm^{-2}; $T = 82$ K (from Barth *et al.* [7]).

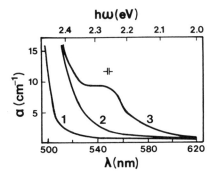

Fig. 5.10 Absorption spectra of control and deformed samples at 77 K. (1) Control sample; (2) sample deformed by 4.8%, polarization of light perpendicular to maximum shear direction; (3) sample deformed by 4.8%, polarization of light parallel to maximum shear direction (from Klassen and Osipyan [11]).

fact that a polarization-dependent absorption band attributable to dislocations is also observed in CdS (Fig. 5.10) [11].

2. Electronic Structure at a Free Surface

The electronic structure of semiconductor surfaces is reviewed by Pandey [12] and Phillips [13]. Hence only the relevant aspects of this subject are noted here. First, surfaces restructure, which shifts atoms from their crystalline equilibrium positions and, therefore, causes bond distortions. Second, given hypothetical atomic positions, a formalism exists [linear combination of atomic orbitals (LCAO) or semiempirical tight binding (SETB) methods] that allows the electronic energy levels to be calculated. Third, the LCAO and SETB methods can be extended (thermochemical calculations) to allow prediction of the lowest-energy state atomic configurations. All these statements should apply equally well to the hypothetical dislocation structures in Fig. 5.2. Potentially, these theoretical techniques might also be applicable to grain boundary structures.

The dangling bonds on free surfaces are quite reactive. Thus, a silicon-free surface exposed to oxygen or hydrogen restructures such that the free surface states originally present are removed from the energy gap. This observation is again consistent with the hypothesized ability of dangling bonds at dislocations to heal by impurity gettering.

3. Electronic Structure of a Grain Boundary

The structure of grain boundaries in metals has long been of interest to metallurgists. Gleiter [14] has reviewed the various theoretical models (i.e., the dislocation, island, coincident site, structural unit, and ledge models). The details of these models are not of interest here. However, he summarizes the metallurgical data and concludes that the atoms at a grain boundary form a periodic array rather than an extended region represented as an amorphous (glassy) structure. This conclusion has been confirmed by photographs taken with scanning transmission electron microscopy (STEM) [15, 16]. These photographs show essentially atomic dimensions. Relevant to this discussion is a STEM study of a germanium grain boundary published by Krivanek et al. [16]. Figure 5.11 shows a photograph from this study. Based on this photograph, Krivanek proposes a model for the relevant grain boundary atomic structure. His proposed model (Fig. 5.11a), the relevant region of his original photograph (Fig. 5.11b), and a blurred image of the model (Fig. 5.11c) are shown here. The agreement between theory and experiment is quite convincing. Close inspection of Krivanek's grain boundary model shows alternate rings of five and seven atoms connecting the two adjacent crystals. No dangling bonds are required but, as in the case of the dislocation, distorted bonds are involved. In fact, the alternating five- and seven-ring structure resembles the dislocation structure in Fig. 5.2c. The electronic structure of this grain boundary, then, can be modeled much as was done for the dislocation except that, for the grain boundary, the defect states arising from dilated bonds will form two-dimensional anisotropic bands.

We now turn to the interaction of the defect states with the bulk crystal, noting that again a depletion width will result as in Fig. 5.5. In this

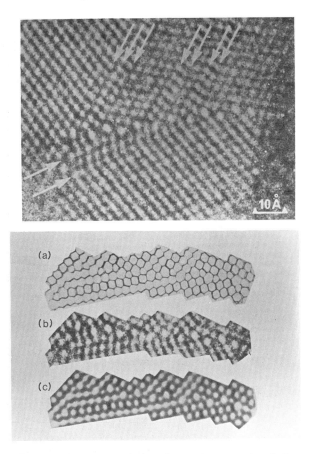

Fig. 5.11 (Top) TEM picture of grain boundary region separating single crystals A and C of germanium. (Bottom) A comparison of the model (a), the TEM picture (b), and the blurred image of the model (c) (from Krivanek *et al.* [16]).

case, however, the depletion width will be planar, and the equations relating the depletion width L and the trapped charge n_+ to the core level depth ϕ and bulk dopant concentration N are

$$\phi = qn_+^2/8\epsilon N \qquad (5.3)$$

$$n_+ = LN \qquad (5.4)$$

In the formation of polycrystalline resistors or polycrystalline photoconductors, the current conduction across the grain boundary potential barriers of height ϕ can determine the device's performance characteristics. This is generally known and has been described by various authors [17, 18]. However, what is not generally known is that the two-dimensional banding of the defect states in the grain boundary leads to high conductivity along

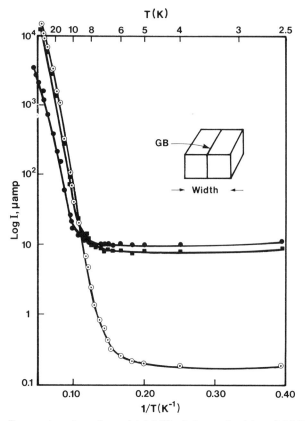

Fig. 5.12 Temperature dependence (at 1.0 V) of the conductivity of 1.0-Ω-cm p-type germanium and of the grain boundary region (sheet conductance): $\odot$—bulk, 0.03 in. wide; $\bullet$—GB, 0.03 in. wide; $\blacksquare$—GB, 0.003 in. wide (from Mataré [19]).

grain boundaries [19]. This has been shown for a germanium bicrystal (Fig. 5.12). At low temperatures, the carriers in the bulk crystal freeze out, and the remaining conductivity results from the partial filling of the grain boundary defect band. At 4 K, the sheet conductivity of the grain boundary ranges from 2000 to 10,000 $\Omega/\square$. This parallel grain boundary conduction will be of importance for thin-film diodes and solar cells, as discussed in Section 5.3.

5.3 GRAIN BOUNDARY EFFECTS IN POLYCRYSTALLINE THIN-FILM DEVICES

Before beginning a discussion of the effects of grain boundaries on thin-film devices, it would be helpful to distinguish between intrinsic and extrinsic grain boundaries. We define an extrinsic grain boundary as a grain boundary where second-phase precipitation occurs or where the doping concentratiom in the adjacent crystalline depletion region is appreciably different from that of the bulk crystal dopant concentration. We define an intrinsic grain boundary as one with its properties controlled by intrinsic effects such as lattice distortions. We will include the healing of dangling bonds in the intrinsic category because, even though impurity gettering can occur, it does not automatically produce second-phase precipitation or doping. In most work on thin-film devices, this distinction is not made. Since grain boundary diffusion can occur quite rapidly [20] and preferentially, and since most thin-film devices are fabricated on foreign substrates, many of the grain boundary effects observed are probably extrinsic [21]. Here we take the point of view that it is first necessary to understand the intrinsic effects. After learning how to control the intrinsic properties of grain boundaries, the extrinsic effects can then be defined and studied.

Polycrystalline devices include resistors, thin-film transistors (TFTs), photoconductors, diodes, and solar cells. The effects of grain boundaries on resistors, TFTs, and photoconductors have been discussed elsewhere. Herein we describe the effect of intrinsic grain boundaries on diode $I-V$ curves. The diode discussion relates to the value of V_{0c} in thin-film solar cells.

We envision a polycrystalline thin-film homojunction diode fabricated by successive depositions of p- and n-type material. We assume that the film grains are columnar. Then the junction will lie perpendicular to the grain boundaries. Figure 5.13 shows an expanded view of a representative region including the $p-n$ junction and an intersecting grain boundary. We have observed that the grain boundary region can be represented as a two-dimensional semiconductor with a reduced band gap. The distorted grain boundary core shifts the conduction band downward by ϕ_c and the valence band upward by ϕ_v. Thus, if the bulk semiconductor energy gap is E_G, the grain boundary will behave as a semiconductor with band gap $E_{GB} = E_G - \phi_c - \phi_v$.

The $I-V$ characteristics of a polycrystalline $p-n$ junction, then, can be modeled in terms of parallel bulk and grain boundary diodes with areas and activation energies A_G (E_G) and A_{GB} (E_{GB}), respectively. The band picture for this model is shown in Fig. 5.13. From this model, it is immediately clear that, although A_{GB} is usually small in comparison to A_G, this factor can be rapidly compensated for by the fact that the activation energy occurs in an exponent and its value is lower for the grain boundary diode.

This model can be carried still further if we assume $I-V$ diode characteristics controlled by generation–recombination mechanisms. In this case, the first-order defect states at the grain boundary (i.e., bond

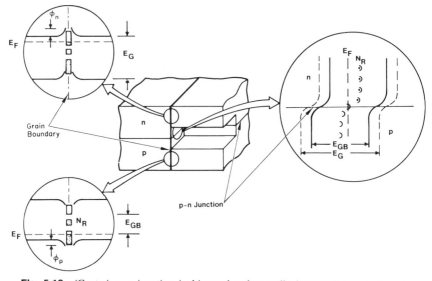

Fig. 5.13 (Center) $p-n$ junction in bicrystal orthogonally intersecting grain boundary. (Left) Schematic representation of energy diagram across grain boundary region in n-type (top) and p-type (bottom) material. (Right) Schematic representation of energy diagram grain boundary material intersecting the $p-n$ junction. Dashed lines refer to band edges of adjacent single crystal.

distortions) lead to the banding states, and the second-order defect states (i.e., dangling bonds and jogs) lead to recombination centers with a hypothetical density N_R and an associated grain boundary minority carrier lifetime τ_{GB}. The grain boundary diode leakage current can then be written

$$I_{GB} = A_{GB}(en_{GB}w/\tau_{GB}) \exp\left[-(E_G - \phi_c - \phi_v)/2kT\right] \qquad (5.5)$$

where w is the diode depletion width. Similarly, the bulk diode leakage current can be written.

$$I_G = A_G(en_G w/\tau_G) \exp(-E_G/2kT) \qquad (5.6)$$

The total leakage current will be

$$I_0 = I_{GB} + I_G \qquad (5.7)$$

For a polycrystalline solar cell, a large leakage current will limit V_{0c}. It is relevant, then, to ask at what grain size the grain boundary leakage current will equal the bulk leakage current. From the above equations and by assuming $I_{GB} = I_G$, we can write

$$A_G = A_{GB}(n_{GB}\tau_G/\tau_{GB}n_G) \exp\left[(\phi_c + \phi_v)/2kT\right] \qquad (5.8)$$

TABLE 5.1

Grain Sizes Below Which Grain Boundaries
Contribute Dominant Cell Leakage Currents[a]

$\phi_c + \phi_v$ (eV)	τ_G/τ_{GB}	Grain size (μm)
0.3	10	4
0.4	10	30
0.5	10	200
0.3	100	40
0.4	100	300
0.5	100	2000

[a] The lifetime ratio τ_G/τ_{GB} and the gap
reduction energy $\phi_c + \phi_v$ are parameters.

Table 5.1 lists the grain sizes below which the grain boundaries will be the dominant contributors to the cell leakage current. We assume for A_{GB} a grain boundary core width of 10 Å. The factor $n_{GB}\tau_G/\tau_{GB}n_G$ and the gap reduction energy $\phi_c + \phi_v$ are treated as parameters. For most semiconductors, we would expect $n_G \sim n_{GB}$ and $\tau_G \gg \tau_{GB}$. Because the gap lowering in the core of the germanium dislocation was 0.22 eV and because germanium is a low-gap semiconductor, we would expect $\phi_c + \phi_v$ to be greater than or equal to 0.3 eV for solar cell materials. Therefore, we conclude that the leakage current in polycrystalline solar cells with grain sizes less than approximately 4 μm is controlled by the grain boundaries. Given the current interest in thin-film solar cells, this implies that the electronic structure of grain boundaries should be an area of future experimental and theoretical study.

5.4 MEASUREMENT OF GRAIN BOUNDARY STRUCTURES

The success in measuring the electronic structure of dislocations was predicated on the ability to prepare nominally identical crystals with and without dislocations. Similarly, measurement of the electronic structure of grain boundaries will require controlled experiments in which films of identical purity are prepared with and without grain boundaries. In practice, this requirement translates to the requirement that polycrystalline films and control single-crystal films be prepared in the same system under nearly identical growth conditions.

Molecular beam epitaxy (MBE) has three distinct advantages in this regard. First, high-purity films can be prepared ($\leqslant$ 1 ppm); second, films can be prepared with low substrate temperatures, thereby alleviating

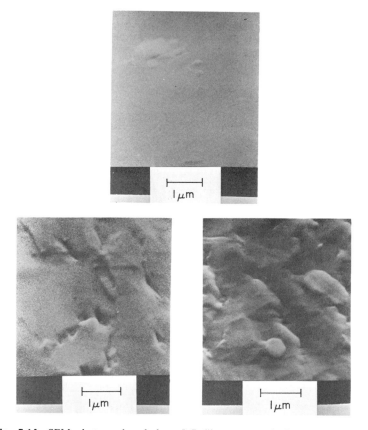

Fig. 5.14 SEM photographs of three InP films grown via MBE at three different substrate temperatures (from McFee *et al.* [22]).

interdiffusion problems; and third, dopants can be used with unity sticking coefficients (i.e., the quantity of dopant incorporated is independent of substrate temperature). This allows the film structure to be varied independent of dopant concentration by simply varying the substrate temperature. McFee *et al.* [22] have grown single-crystal and polycrystalline films of InP with MBE, and their work illustrates the utility of MBE in this regard. They studied film surface morphology using scanning electron microscopy (SEM) (Fig. 5.14), film electrical properties using Hall measurements (Fig. 5.15), and film photoluminescence spectra (Fig. 5.16) as a function of the substrate temperature during film growth. At $\approx 315°C$, the film electrical properties show an abrupt transition, the SEM photograph shows a morphological change, and a lower-energy photoluminescence peak develops.

Interesting work with polycrystalline silicon films prepared with con-

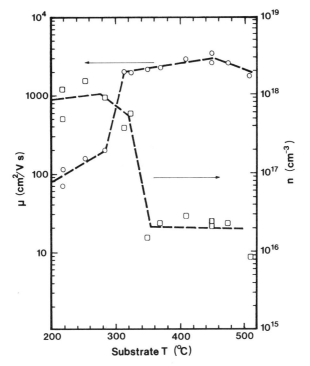

Fig. 5.15 Carrier concentration at 300 K ($n \equiv N_d - N_a$) ($\square$) and Hall mobility (μ) (O) of ($\overline{1}11$) InP epitaxial layers as a function of growth temperature.

trolled dopant concentrations has recently been reported by Seto [23]. He grew nominally undoped polycrystalline films by CVD. Unfortunately, he did not provide reference single-crystal data. However, by ion implantation and subsequent nitrogen gas anneals, he was able to vary the film dopant concentration over three orders of magnitude with basically identical processing, thereby ensuring that the structural quality of the variously doped films would be identical.

Seto's film measurements are of particular interest here. By making Hall measurements as a function of temperature, he arrived at activation energies for free carrier transport as a function of film dopant concentration. His data for p-type polysilicon are shown in Fig. 5.17a. He also plotted film mobility versus dopant concentration (Fig. 5.17b). The mobility data show that there are three transport regions. First, since his film grain size was only 200 Å, for carrier concentrations below $1 \times 10^{18}/\mathrm{cm}^3$, the film grains are completely depleted. Second, at a concentration of approximately $2 \times 10^{18}/\mathrm{cm}^3$, the depletion width is slightly smaller than the grain size, and the grain boundary band begins to fill. In this range, the

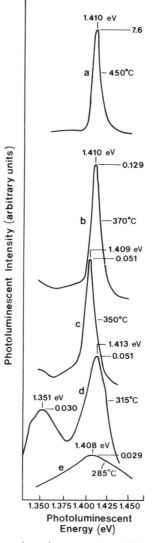

Fig. 5.16 Photoluminescent intensity spectrum at 77 K of ($\overline{111}$) InP epitaxial layers grown at various temperatures indicated on curves. Intensity scale is linear, but curves are displaced vertically for clarity so they do not overlap. Peak heights are indicated on curves. Argon ion laser line at 4880 Å used for excitation (from McFee *et al.* [22]).

mobility is at a minimum, because the grain boundary barrier height is at a maximum relative to the bulk of the grain. Third, at higher doping concentrations, the defect band begins to fill, leading to a reduced barrier height. The barrier also becomes narrower, and tunneling can occur. This

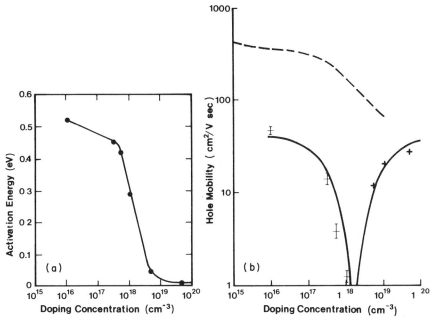

Fig. 5.17 (a) Experimental activation energy as a function of doping concentration (from Seto [23]). (b) Room-temperature hole Hall mobility versus doping concentration. The experimental result (+) is plotted together with the theoretical solid curve. The broken line is for single-crystal silicon (from Seto [23]).

Hall measurement technique can be used to measure the grain boundary density of states. This would be applicable in the range in which the depletion width is smaller than the grain size, but yet too large for tunneling. Since Seto's grain size was only 200 Å, this region is very narrow. However, an approximate value of ϕ_v can be obtained from the activation energy at the carrier concentration where μ is minimal (i.e., $p \sim 2 \times 10^{18}/\mathrm{cm}^3$, $\phi_v \sim 0.2$–0.3 eV). The main point is that this Hall measurement procedure for n- and p-type polycrystalline films can lead to a density of states distribution at the grain boundaries and, thereby, to values of ϕ_c and ϕ_v.

After having deposited polycrystalline films under controlled conditions (MBE) and after making transverse Hall measurements to determine the values of ϕ_c and ϕ_v, the next logical measurement step would be to fabricate a polycrystalline n–p homojunction using MBE and to look at diode I–V characteristics. Hopefully, for small-grained films, the forward-current characteristic would show a diode activation energy of $E_G - \phi_c - \phi_v$, which would verify the model. This would assume a known diode quality factor. In Eq. (5.5) we assumed a value of 2.

The above discussion relates to the measurement of the electronic structure of grain boundaries. STEM micrographs of a variety of grain boundary types in a variety of semiconductor materials will also be of value in terms of the atomic structure of grain boundaries.

5.5 GRAIN BOUNDARY PASSIVATION

Our model of the origins of the electronic states at grain boundaries suggests some means of passivation. First, with regard to potential dangling bonds at grain boundaries, heat-treating in hydrogen might be quite effective in hydrogenating these bonds. In this regard, Andrews [24] has reported experiments similar to those of Seto, but with phosphorus-implanted polysilicon. He observed that the film resistivity decreased by as much as two orders of magnitude after heat treatment in hydrogen at 450°C. Unfortunately, he did not attempt heat treatment in other ambients.

The problem of distorted (dilated) bonds is probably the first-order problem. The solution might be to replace the atoms at the grain boundaries with atoms forming stronger covalent bonds (e.g., C in Si or P in GaAs). This could be viewed as cladding the grain with a higher band gap material or, alternately, as simply strengthening the distorted bond (see Fig. 5.18). Elements forming strong covalent bonds are, for example, H, B,

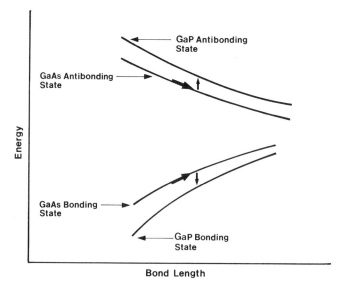

Fig. 5.18 If dilation of GaAs bonds pulls states out of the conduction and valance bonds, perhaps replacing the GaAs bond at the grain boundary with a stronger bond will move the defect states back into the bands.

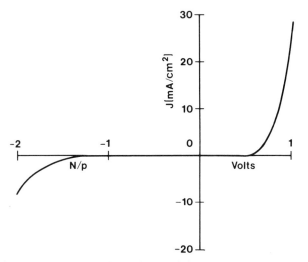

Fig. 5.19 Current density–voltage characteristics for an *n*-type AlAs–*p*-type GaAs polycrystalline thin-film diode (from Johnston *et al.* [25]).

C, P, and S. One might envision carrying out such a passivation either by heat treatment (in which case preferential grain boundary diffusion would be relied on) or simply by fabricating semiconductor alloy materials (in which case preferential segregation at grain boundaries would be relied on).

Johnston has presented data on AlAs–GaAs polycrystalline solar cells [25]. Although he was not able to produce a functional transparent contact, his dark-state (metal-contacted) diode curves look reasonably good (Fig. 5.19). Since his AlAs layers were deposited at 1000°C on top of the GaAs polycrystalline films, it is tempting to conclude that the GaAs grain boundaries were passivated by a cladding of AlAs.

An alternative technique of grain boundary passivation is to fabricate an extrinsic, highly resistive grain boundary by diffusing a deep-level impurity into the grain boundary region. If the concentration of the impurity were high enough, it could deplete the grain boundary banding states; if it diffused far enough into the bulk adjacent to the grain boundary, it would not produce banding itself. This may be the grain boundary passivation mechanism in the $CdS–Cu_2S$ system via Cu diffusion into CdS grain boundary regions.

5.6 CONCLUSIONS

Based on an experimentally verified model for the electronic structure of dislocations, we have derived a model for the intrinsic properties of grain boundaries in polycrystalline semiconductor thin films. Although the model is speculative, it is consistent with a variety of experimental observations. We have described experimental means by which this model can be verified and extended. If correct, the model may have some impact in the design of high-efficiency, thin-film solar cells.

REFERENCES

1. C. Schiller and M. Boulov, *Philips Tech. Rev.* **35**, 239 (1975).
2. V. A. Phillips and R. Wagner, *J. Appl. Phys.* **44**, 4252 (1973).
3. E. Kamieniecki, *J. Phys. C Solid State Phys.* **9**, 1211 (1976).
4. J. Hornstra, *J. Phys. Chem. Solids* **5**, 129 (1958).
5. F. O. Wöhler, H. Alexander, and W. Sander, *J. Phys. Chem. Solids* **31**, 1381 (1970).
6. J. C. Slater, "Quantum Theory of Molecules and Solids." McGraw Hill, New York, 1965.
7. W. Barth, K. Elsaesser, and W. Güth, *Phys. Status Solidi (a)* **34**, 153 (1976).
8. W. Barth, M. Bettini, and U. Ostertag, *Phys. Status Solidi (a)* **3**, K177 (1970).
9. R. Labusch and W. Schroter, *Inst. Phys. Conf. Ser. No. 23, London, England*, p. 56 (1975).
10. S. Mantovani, U. DelPennino, and E. Mazzega, *Phys. Status Solidi (a)* **35**, 451 (1976).
11. N. V. Klassen and Y. A. Osipyan, *Sov. Phys. – Solid State* **14**, 3094 (1973).
12. K. C. Pandey, *J. Vac. Sci. Technol.* **15**, 440 (1978).
13. J. C. Phillips, *Surface Sci.* **53**, 474 (1975).
14. H. Gleiter, *Phys. Status Solidi (b)* **45**, 9 (1971).
15. J. W. Matthews and W. M. Stobbs, *Phil. Mag.* **36**, 373 (1977).
16. O. L. Krivanek, S. Isoda, and K. Kobayashi, *Phil. Mag.* **36**, 931 (1977).
17. R. H. Bube, *Annu. Rev. Mater. Sci.* **5**, 201 (1975).
18. J. C. Anderson, *Vacuum* **27**, 263 (1977).
19. H. F. Mataré, "Defect Electronics in Semiconductors," p. 272. Wiley (Interscience), New York, 1971.
20. K. Nakamura and M. Kamoshida, *J. Appl. Phys.* **48**, 5349 (1977).
21. L. M. Fraas, *J. Appl. Phys.* **49**, 871 (1978).
22. J. H. McFee, B. I. Miller, and K. J. Bachmann, *J. Electrochem. Soc.* **124**, 259 (1977).
23. J. Y. W. Seto, *J. Appl. Phys.* **46**, 5247 (1975).
24. J. M. Andrews, *Electron. Mater. Conf., 20th, Santa Barbara, California, 28–30 June* (1978).
25. W. D. Johnston, Jr., and W. M. Callahan, *J. Electrochem. Soc.* **125**, 979 (1978).

6 | Amorphous Thin-Film Devices

D. E. CARLSON

RCA Laboratories
Princeton, New Jersey

6.1 INTRODUCTION

The field of amorphous semiconductors has attracted considerable interest in recent years with the development of electronic switching and memory devices, optical storage systems, and solar cells. The main objective of this chapter is to review recent developments in thin-film devices fabricated from hydrogenated amorphous silicon (a-Si:H). However, a brief historical review of amorphous semiconductor devices in general is given in Section 6.2.

175

POLYCRYSTALLINE AND AMORPHOUS
THIN FILMS AND DEVICES

In Section 6.3 we discuss the methods of producing a-Si:H as well as the properties of the material. In Section 6.4 we review the experimental thin-film devices that have been fabricated from a-Si:H, including photo-conductivity cells, diodes, solar cells, photodetectors, optical storage systems, and transistors. Finally, we conclude the chapter with some conjectures on future directions for device applications.

6.2 HISTORICAL REVIEW OF AMORPHOUS SEMICONDUCTOR DEVICES

The first major application of amorphous semiconductors was in the field of Xerography. The concept was first proposed by Chester F. Carlson in 1938, but commercialization of the process did not occur until two decades later. This process utilizes the photoconductivity of certain high-resistivity, amorphous semiconductors containing selenium. The amorphous film is generally ~ 50 μm thick and is charged positively by a corona discharge. Exposing the film to light generates electron–hole pairs that are separated by the electric field in the film, and the surface charge is neutralized. Negatively charged toner particles (carbon black in a low-melting plastic) are attracted to those regions of the film that were not exposed to light, and another corona discharge is used to transfer the toner to a sheet of paper where the image is fixed by heating. For a more detailed discussion of the Xerographic process see Mott and Davis [1] or Dessauer and Clark [2].

Recently, Xerox Corporation has developed a color copier that employs a three-layer structure of different amorphous semiconductors (chalcogenide glasses that are sensitive to different portions of the visible spectrum). The photoconductive properties of chalcogenide glasses are also utilized in an image pickup tube (the Saticon) developed by Hitachi and the Japan Broadcasting Corporation for use in small color television cameras. The active element of the tube is a graded amorphous alloy of selenium, arsenic, and tellurium [3].

The phenomenon of switching in amorphous semiconductors was discovered by Stanford R. Ovshinsky in 1958; this unpublished work was cited by Young [4] and later reviewed by Ovshinsky [5]. Switching refers to a rapid change in the electrical conductivity of a semiconductor when the applied field reaches a critical value. In chalcogenide glasses the switching occurs at a field on the order of 10^5 V/cm and in a time on the order of nanoseconds [6]. The electric conductivity can change by as much as a factor of $\sim 10^6$. The mechanism for switching appears to be electronic in nature (see, e.g., Holmberg and Shaw [6]), but some experimental work indicates that thermal effects may play a role in some cases [7, 8].

Memory switching has also been observed in amorphous semiconductors [9]. In this type of switching, the high-conductivity state is retained even after the applied voltage is removed. The low-conductivity state is reestablished by applying a strong current pulse. The memory switching is due to a reversible transition between the amorphous and the crystalline states. Several companies are developing an electrically alterable, read-only computer memory by using the two different states to store information.

Some chalcogenide glasses exhibit reversible optical changes that can be induced by light [10]. This optical memory switching is due to a reversible transition between the amorphous and crystalline states. Other chalcogenide glasses exhibit irreversible changes in their optical properties upon exposure to intense light. In some cases, the mechanism for changing the structure of the glasses appears to be purely thermal [11], while in others there is evidence for photoenhanced crystallization [12]. The general area of imaging with amorphous semiconductors appears to hold great promise for the future.

Rectifying junctions have been fabricated by sputtering amorphous silicon (no hydrogen present) onto crystalline silicon substrates [13]. However, most of the space charge region exists in the crystalline silicon because of the high density of localized states in the amorphous silicon. Thin-film transistors (TFTs) were constructed using amorphous silicon deposited by electron beam evaporation [14]. The amorphous silicon (~ 600 Å thick) was deposited on top of ~ 3000 Å of a thermal oxide grown on an n^+ crystalline silicon gate. These devices exhibited large threshold gate voltages (> 50 V) because of the large density of localized states ($\sim 10^{20}$ cm^{-3} eV^{-1}) in evaporated amorphous silicon.

Hydrogenated amorphous silicon (a-Si:H) was first investigated by Chittick *et al.* [15]. They observed that amorphous silicon produced in a silane (SiH$_4$) glow discharge exhibited a large photoconductive effect as compared to evaporated or sputtered amorphous silicon (hydrogen-free). Electronic devices using a-Si:H were first made at RCA Laboratories in 1974 [16], and the development of these devices is discussed in Section 6.4.

6.3 HYDROGENATED AMORPHOUS SILICON

In this section we discuss the techniques that have been used to deposit a-Si:H, the role of hydrogen, and the optical and electrical properties of the material.

1. Deposition Techniques

Hydrogenated amorphous silicon was first deposited from a glow discharge in silane (SiH_4) generated by an external rf coil [15]. Most of the early researchers investigating a-Si:H used this technique. These rf electrodeless discharge systems usually operate within a frequency range of 0.5–13.5 MHz and at SiH_4 pressures of ~0.1–2.0 torr. The discharge chambers are generally small (on the order of 7 cm in diameter), and the samples are usually positioned horizontally on a heated pedestal. The best quality films are obtained with substrate temperatures in the range of 200°–400°C. The SiH_4 flow rates are typically in the range of 0.2–5.0 standard cm^3/min (sccm), and the deposition rates are usually in the range of 100–1000 Å/min. The deposition rate increases rapidly with increasing power and pressure, but the film quality starts to degrade at high deposition rates. At high pressures, the degradation appears to be due to increasing polymerization or gas-phase nucleation in the discharge [17], while at high powers energetic electrons and ions may cause bombardment damage as well as increased polymerization [18]. These rf electrodeless discharge systems have the advantage of external electrodes, so that contamination of the films by sputtering is minimized; however, these systems are relatively small, and the uniformity of the film is generally poor.

Radio frequency capacitive discharge systems employ parallel plate electrodes inside the discharge chamber and are similar to the systems used for sputter-deposition [19]. The capacitive discharge systems generally operate at 13.5 MHz and at SiH_4 pressures in the range of 5–250 mtorr. The film uniformity is excellent, and some systems can accommodate substrates over 1 ft^2 in area. The SiH_4 flow rates are typically 10–50 sccm and depend on the size of the system as well as on the pressure and power. The power densities are usually in the range of 0.1–2.0 w/cm^2, and the deposition rates are in the range of 50–500 Å/min. These systems can also be used to sputter-deposit contacts onto a-Si:H films [20].

Deposit of a-Si:H has also been done by means of a dc glow discharge in SiH_4 [16]. Deposition rates in the range of 0.1–1.0 μm/min can be obtained by varying the current density to a cathodic substrate from 0.2 to 2.0 mA/cm^2 in ~1.0 torr of SiH_4. While films can also be deposited on an anodic substrate, the deposition rate is roughly an order of magnitude slower and the film quality is generally poor. Cathodic films are bombarded by energetic positive ions during the deposition, and bombardment damage can degrade the film quality at high voltages and low pressures. Film uniformity is good if the substrate dimensions are much larger than the Crookes dark space (a dark region in the glow discharge near the cathode).

Bombardment damage can be reduced by electrically isolating the substrate and by placing a cathodic screen between the anode and the substrate [16]. The deposition rate is less than that for a cathodic substrate and depends on the SiH_4 pressure, the applied voltage, and the spacing between the cathodic screen and the substrate. The spacing should be on the order of the dark space that occurs near the cathode. If the screen–substrate spacing is too small, some energetic ions will induce damage in the growing film, and if the spacing is too large, gas-phase polymerization or nucleation will occur before deposition. One disadvantage of this technique is that the cathodic screen tends to flake when film thicknesses approach a few microns.

Another technique that has been used to deposit a-Si:H is sputtering in an atmosphere of Ar and H_2 [21]. In this approach a polycrystalline silicon target is sputtered in an rf capacitive discharge system. The substrate is placed on a heated counterelectrode located opposite the target electrode. This technique has the advantage of being able to control the hydrogen content of the films by varying the partial pressure of H_2. However, preliminary results obtained at RCA Laboratories indicate that the film quality is inferior to that obtained in a SiH_4 discharge.

Other approaches for producing a-Si:H that are currently under investigation include the evaporation of amorphous silicon in the presence of H_2 [22] and the hydrogenation of evaporated and sputtered films [23]. In the latter case, films deposited by evaporation or sputtering in a hydrogen-free environment are exposed to a hydrogen glow discharge at temperatures of $\sim 300°C$. However, the quality of these films with respect to device applications has not yet been determined.

2. The Role of Hydrogen

Lewis *et al.* [24] were the first to demonstrate that hydrogen could compensate dangling bonds in tetrahedrally bonded amorphous semiconductors. They showed that the density of gap states in sputtered amorphous germanium was reduced when H_2 was added to the Ar sputtering atmosphere. Sakurai and Hagstrum [25] showed that atomic hydrogen could compensate dangling bonds on the surface of crystalline silicon. The effect of hydrogenation is to remove surface states near the Fermi level and to create new bonding states that are several electron volts below the Fermi level.

Triska *et al.* [26] were the first to show that discharge-produced amorphous silicon contained considerable amounts of hydrogen. Subsequently, other investigators have shown that the hydrogen content is in the range of 10–50 at.% and that most of the hydrogen is bonded to silicon atoms [27, 28]. For a-Si:H films deposited at substrate temperatures above

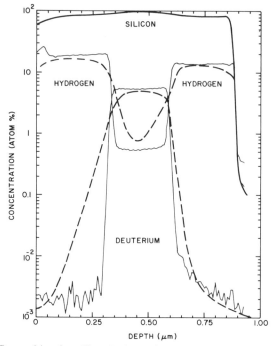

Fig. 6.1 Compositional profiles of a layered sample (made in discharges of either SiH$_4$ or SiD$_4$) before (solid curves) and after (dashed curves) 20 min at $\sim 330°$C. A secondary ion mass spectrometer was used to obtain the profiles.

$\sim 200°$C, most of the hydrogen is present in the monohydride form. Films deposited at lower substrate temperatures contain significant amounts of dihydride and possibly trihydride groups, as well as short polymer chains. These low-temperature films exhibit high defect densities that appear to be due mainly to the presence of short polymer chains [18].

As the substrate temperature T_s increases above $\sim 350°$C, the defect density starts to increase because of the evolution of hydrogen from the film and the concurrent generation of dangling bonds [29]. Carlson and Magee [30] showed that hydrogen diffuses in a-Si:H with an activation energy of ~ 1.5 eV, and thus the evolution of hydrogen should not be a problem for devices operating at temperatures $< 100°$C. However, the diffusion of hydrogen as well as other properties (see Section 6.2.3) depends on the hydrogen content of the film. This behavior is illustrated in Fig. 6.1, which shows compositional profiles of hydrogen and deuterium in a layered film before and after a heat treatment at $\sim 330°$C. The data indicate that the diffusion coefficient increases as the hydrogen content increases [i.e., the diffusion coefficient of deuterium in the top layer (~ 19

at. % H) is a factor of 5 larger than that in the bottom layer (~ 14 at. % H)].

Recent electron diffraction studies indicate that discharge-produced a-Si:H films ($T_s \sim 205°C$) possess a greater degree of short-range tetrahedral ordering than evaporated films [31]. Also, Tanielian *et al.* [32] found that the density of a-Si:H films increased from 1.92 to 2.17 g/cm³ as the H to Si ratio decreased from 0.35 to 0.13. These results suggest that the hydrogen in a-Si:H films ($T_s > 200°C$) is incorporated in the form of compensated microvoids. Monovacancies containing four hydrogen atoms or divacancies containing six hydrogen atoms should be relatively immobile and should not give rise to localized states in the gap [33]. Even larger compensated microvoids may be present in a-Si:H films, and the size and concentration of these microvoids probably depends strongly on the deposition conditions.

For an rf electrodeless discharge, the hydrogen content increases as the substrate temperature decreases, as the SiH_4 pressure decreases, and as the rf power increases [34]. Figure 6.2 shows a compositional profile of the hydrogen content in a layered film deposited at six different power settings ($T_s \simeq 305°C$). The variation of the hydrogen content from ~ 16 to ~ 52 at. % causes variations in the film properties, but since relatively good photovoltaic characteristics can be obtained over this entire range, the concentration of hydrogen is not a critical parameter in obtaining good electronic properties.

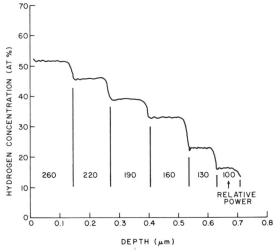

Fig. 6.2 Hydrogen concentration as a function of depth for a sample with layers deposited at relative power levels ranging from 100 to 260. A setting of 100 was the threshold voltage for 0.8 torr of SiH_4.

3. Optical and Electrical Properties

The addition of bonded hydrogen to the amorphous silicon structure causes the top of the valence band to move downward, and thus the energy gap increases [35]. The variation of the energy gap with hydrogen content is reflected in the optical absorption data shown in Fig. 6.3. As the substrate temperature decreases, the hydrogen content increases, and the increasing optical gap causes a decrease in the absorption coefficient α [28]. The optical gap is determined by plotting $(\alpha h\nu)^{1/2}$ as a function of the photon energy $h\nu$ [36]. The variation in the optical gap causes corresponding variations in the photovoltaic properties, with the short-circuit current decreasing and the open-circuit voltage increasing as the gap increases [34]. The refractive index of a-Si:H is close to that of crystalline silicon and decreases only slightly ($\sim 5\%$) as the hydrogen content increases from ~ 10 to ~ 50 at. % [28].

For substrate temperatures in the range of $200°–400°C$, the optical gap is generally in the range of 1.8–1.6 eV. As shown in Fig. 6.3, optical absorption coefficients on the order of 10^3 cm^{-1} are observed for photon energies less than the band gap. This absorption is not directly related to the hydrogen content but appears to be due to localized states in the gap

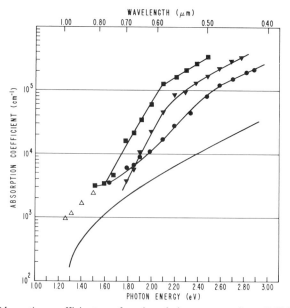

Fig. 6.3 Absorption coefficient as a function of photon energy for a-Si:H films produced in a dc discharge at different substrate temperatures. Circles, 210°C; solid triangles, 325°C; squares, 415°C; solid line, crystalline silicon; open triangles, 325°C, 3 μm-thick film.

[28]. Photoconductivity due to these states can also be observed at photon energies less than the band gap [37].

The dark resistivity of undoped a-Si:H can vary from $\gtrsim 10^{11}$ Ω cm for deposition at $\sim 100°C$ to $\lesssim 10^5$ Ω cm for $T_s \simeq 550°C$ [15]. Recently, Staebler and Wronski [38] reported that large, reversible conductivity changes could occur in a-Si:H films depending on their thermal history and their exposure to light. The dark resistivity of an annealed film can be increased by as much as four orders of magnitude by exposure to light, but the original resistivity can be restored by annealing the film for several minutes at $> 150°C$.

The behavior of the dark conductivity and photoconductivity of undoped a-Si:H is consistent with free-carrier transport over a wide temperature range [37, 39]. The dark conductivity can be described by the expression

$$\sigma = \sigma_0 \exp(-E_a/kT) = \sigma_0 \exp\{-[E_c - E_f(T)]/kT\} \qquad (6.1)$$

where the activation energy E_a can vary from ~ 0.2 to ~ 0.8 eV for undoped a-Si:H [15]; $E_c - E_f(T)$ is the position of the Fermi level with respect to the bottom of the conduction band. The preexponential term σ_0 in Eq. (6.1) exhibits a dependence on E_a that has been attributed mainly to a statistical temperature shift in the Fermi level [40].

The conduction in a-Si:H becomes nonohmic when the average electric field becomes larger than $\sim 10^4$ V/cm. An investigation of the high field conduction as a function of voltage, temperature, and film thickness indicates that the Poole–Frenkel effect (see e.g., Simmons [41]) is the most likely mechanism [42].

The photoconductivity of undoped a-Si:H can be described by the recombination and trapping kinetics discussed by Rose [43]. In general, the photoconductivity for monochromatic light can be expressed as

$$\sigma_p = K_1 N \mu_n \tau_n \qquad (6.2)$$

where N is the generation efficiency of free carriers ($N \simeq 1$ for $h\nu$ greater than the band gap), μ_n the extended-state electron mobility and τ_n the electron lifetime; K_1 is a factor that depends on the photon flux absorbed in the film. Using Eq. (6.2), Zanzucchi *et al.* [28] estimated that electron lifetimes can be as long as 10^{-4} s for a-Si:H films deposited at $\sim 300°C$. The hole lifetime for lightly doped p-type films is roughly an order of magnitude less than the electron lifetime in undoped a-Si:H (which is slightly n-type). Minority carrier lifetimes have only been estimated indirectly from the photovoltaic characteristics and appear to be on the order of 10^{-8} s.

The dependence of the photoconductivity on light intensity F is given by the expression

$$\sigma_p = K_2 F^\gamma \tag{6.3}$$

where K_2 is a factor independent of light intensity. In annealed films at light intensities greater than $\sim 10^{-3}$ AM1 (AM1 = 1 sun $\simeq 100$ mW/cm^2), the exponential factor γ is usually equal to ~ 0.5, indicating a bimolecular type of recombination. This behavior is consistent with trap-dominated photoconductivity where shallow traps ~ 0.2 eV below the conduction band are dominant [44]. Some a-Si:H films exhibit a γ equal to ~ 0.75 over a variation in light intensity of several orders of magnitude, and this behavior is attributed to an exponential distribution of gap states. Prolonged exposure to light changes γ to ~ 0.9 or close to monomolecular recombination, suggesting that centers near midgap then dominate.

Drift mobility measurements indicate that the extended-state mobility is ~ 1–10 cm^2/V s for electrons and that the traps are located ~ 0.2 eV below the conduction band [39]. Moore [45] has performed similar measurements in lightly doped p-type material and found that the hole-trapping state is ~ 0.35 eV above the valence band.

The density and distribution of gap states has been measured by the field-effect technique, and the average density is $\sim 10^{17}$–10^{18} cm^{-3} eV^{-1} [46]. However, these measurements neglect the effect of surface states, which are on the order of 10^{13} cm^{-2} eV^{-1} [47, 48]. Thus, the average density of gap states may be closer to 10^{15}–10^{16} cm^{-3} eV^{-1}, which is in agreement with the space charge densities measured in diodes [49] and the measured density of spin centers [29]. The density of gap states exhibits a strong dependence on substrate temperature, and measurements of both the electron spin density [29] and the photoluminescence [50] show that a minimum density of defects is obtained when the substrate temperature is $\sim 330°$C.

Chittick et al. [15] first attempted to dope a-Si:H by adding small quantities of PH$_3$ to the SiH$_4$ discharge atmosphere, but their results were not convincing. In 1974 work at RCA Laboratories showed that PH$_3$ and B$_2$H$_6$ could be used as dopant sources in SiH$_4$ discharges to produce n- and p-type films, respectively [16]. Similar studies were performed independently at the University of Dundee [51]. Some of the phosphorus and boron atoms apparently go into substitutional sites similar to those in crystalline silicon and thus give rise to shallow donor or acceptor states, respectively. However, doping also appears to generate new states near midgap, as evidenced by the reduction in the photoluminescence intensity [52] and in the solar cell collection efficiency [53].

Doping of a-Si:H with boron or phosphorus can change the conductivity over 10 orders of magnitude, and conductivities of $\lesssim 10^{-2}$ $(\Omega\,cm)^{-1}$ can be obtained by adding a few percent of PH_3 or B_2H_6 to an SiH_4 discharge. Spear and LeComber [54] estimated that approximately one-third of the phosphorus atoms act as shallow donors. However, this analysis depended on the field-effect measurements mentioned earlier, and the inclusion of surface states would lower the estimate of the donor concentration. Preliminary measurements of the space charge in lightly doped diodes at RCA Laboratories indicate that the donor concentration is more than an order of magnitude less than the concentration of phosphorus atoms. Possibly many of the dopant atoms are neutralized by hydrogen.

Drift mobility measurements in phosphorus-doped a-Si:H indicate that conduction can occur in either or both of two paths; for undoped and lightly doped samples, extended-state conduction dominates, but for heavily doped samples, donor-band transport dominates [55]. Hall effect measurements have also been interpreted in terms of the same model [56]. Thermopower measurements on doped a-Si:H films also indicate that the dominant contribution for conduction is in the extended states for undoped and very lightly doped samples. As the doping level is increased, the conduction path changes first to hopping in a state ~ 0.2 eV below the conduction band and then to donor-band conduction ~ 0.13 eV below E_c [57].

Many uncertainties still exist in interpreting some of the experimental results and, moreover, some properties of a-Si:H have not yet been determined. The density and distribution of gap states (and tail states) has probably not been accurately determined, since surface states were neglected as mentioned earlier. Minority carrier lifetimes and diffusion lengths have not yet been accurately measured. The mobility of holes in the valence band is still unknown. The effects of many common impurities and defects in a-Si:H have not been determined. Although an exciton model has been proposed to explain the photoluminescence data [58], there are difficulties with this model [59]. Another model proposed to explain some of the photoluminescence data is based on a field-quenching effect associated with ionized donors and acceptors [59]. There remains the question of whether small-polaron conduction occurs in this material, since some recent results in *p*-type films indicate such a possibility [60]. The mechanism for light-induced conductivity changes is still unclear, and some recent results indicate that surface charges may influence the measured conductivity [61].

We conclude this section by voicing optimism that many of these uncertainties will be resolved in the next few years.

6.4 THIN-FILM DEVICES

We now review recent developments in thin-film devices fabricated from a-Si:H.

1. Photoconductivity Cells

A photoconductivity cell or photocell is a simple device whose resistance changes upon exposure to light of certain wavelengths. These devices are used as light-sensing elements in certain types of light meters and light-activated switches.

Photoconductivity cells are characterized by operational parameters such as spectral response, gain, and decay time. The spectral response depends mainly on the optical absorption properties and the thickness of the semiconducting film. Provided the contacts are ohmic and the electron photoconduction dominates, the gain G is given by

$$G = \mu_n \tau_n V / l^2 \qquad (6.4)$$

where μ_n is the extended-state mobility, τ_n the electron lifetime, V the applied voltage, and l the electrode spacing [43]. For semiconductors with large densities of traps, the photocurrent decay time τ_0 is given by

$$\tau_0 \simeq (n_t / n) \tau_n \qquad (6.5)$$

where n_t is the density of trapped electrons and n is the density of free electrons. From Eqs. (6.4) and (6.5) one can see that a long electron lifetime implies a large gain but also a long decay time.

Photoconductivity cells of a-Si:H have been constructed by first depositing a four-stripe electrode pattern on SiO_2 glass and then depositing the a-Si:H film [28]. Metals such as Mo, Nb, Ta, Ti, and Sb often make reasonably good ohmic contacts to undoped or n-type a-Si:H when the a-Si:H is deposited on top of them. A good contact can be ensured by depositing a thin layer of doped a-Si:H on top of the metal electrodes before depositing undoped a-Si:H. The four-stripe electrode pattern allows one to determine whether the contacts are ohmic or not.

The spectral sensitivity of an a-Si:H photoconductivity cell is shown in Fig. 6.4 for an undoped film deposited in a dc discharge at $\sim 330°C$. The cell consisted of ~ 0.7 μm of a-Si:H deposited on metal electrodes 0.8 cm long and 0.1 cm apart. The photocurrent was normalized for an incident photon flux of 10^{14} cm^{-2} s^{-1} at each wavelength. The decrease in photocurrent at long wavelengths is due to the decrease in optical absorption (see Fig. 6.3). The decrease at short wavelengths is probably due to surface recombination, since the blue light is absorbed within a few hundred angstroms of the surface.

This type of cell exhibited gains of ~ 10 and decay times of $\sim 10^{-2}$ s for photon fluxes of $\sim 10^{-13}$ cm^{-2} s^{-1} (after prolonged exposures to light).

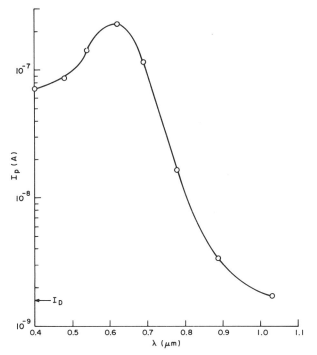

Fig. 6.4 The photocurrent as a function of wavelength for an a-Si:H photoconductivity cell.

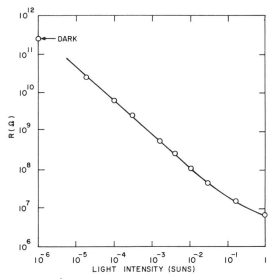

Fig. 6.5 Resistance of an a-Si:H photoconductivity cell as a function of light intensity (simulated sunlight).

The decay time decreases linearly with increasing light intensity for films in a "light-soaked" state, and as the square root of the intensity for films in an annealed state [42]. The resistance of such a photoconductivity cell can change by more than four orders of magnitude when the cell is taken from the dark into sunlight (see Fig. 6.5). These devices can operate up to fields of $\sim 10^4$ V/cm before the resistance becomes nonohmic.

One can fabricate a low-resistance photoconductivity cell by using a-Si:H in a sandwich configuration with thin n^+ layers next to each electrode. The top electrode must be semitransparent, so the structure is more complicated than a simple two-stripe configuration. In the sandwich structure, the resistance can vary from $\sim 10^4$ to ~ 1 Ω when the cell is taken from the dark into sunlight.

2. Diodes

A diode or rectifier is an electronic device that passes current in one direction only. Actually, all diodes will allow a small current to flow in the reverse direction, but this is usually negligible. Generally, the $I-V$ characteristics of a diode can be described by the expression

$$J = J_0\left[\exp(qV/nkT) - 1\right] \tag{6.6}$$

where J_0 is the reverse saturation current density and n the diode quality factor. For an ideal diode $n = 1$, but n can be greater than unity if recombination occurs in the junction or if interface states are present, as in metal insulator semiconductor (MIS) devices [62]. The derivation of Eq. (6.6) is based on the formation of a potential barrier or space charge region within a semiconductor either by selective doping of the semiconductor or by forming a contact with a conductive material of different electrochemical potential [63].

A. STRUCTURES

Since the structures used to make a-Si:H diodes are similar to those used for solar cells, we shall defer a detailed discussion of the structures until Section 6.4.3. One significant difference between the structures is that a simple rectifier does not need a transparent electrode for one contact as in the case of solar cells. Thus, both contacts may be opaque conductors such as metal foils and thick metal films. Another difference is that the diode structure may use doped layers that are more heavily doped and thicker than those used in solar cells.

Common structures that can be used as rectifiers are $p-n$, $p-i-n$, Schottky-barrier, and MIS junctions, as well as heterojunctions and electrolytic junctions. While a-Si:H diodes have been made in all these forms, the most successful have been the Schottky-barrier, MIS, and $p-i-n$ structures.

B. ELECTRICAL CHARACTERISTICS

Figure 6.6 shows the forward-bias $I-V$ curves of a Pt Schottky-barrier diode, an MIS diode, and a $p-i-n$ diode; the reverse bias is also shown for the Pt Schottky-barrier diode. This latter device is clearly the superior rectifier with $n \simeq 1.15$ and $J_0 \simeq 10^{-11}$ A/cm^2. Most Schottky-barrier devices exhibit near-ideal diode behavior, and in general J_0 decreases as the work function of the Schottky barrier metal increases [64]. The MIS device exhibits a diode quality factor of ~ 1.57 as a result of recombination at interface states, and $n \simeq 2.6$ for the $p-i-n$ diode, possibly because of recombination at both interface states and in the p-layer.

Diodes of a-Si:H become resistance-limited at a current density of $\sim 10^{-4}$ A/cm^2 because of the resistance of the quasi-neutral region. As the forward bias is increased, the depletion region decreases, and the thickness of the quasi-neutral region approaches the film thickness. For a 1-μm-thick film and a resistivity of 10^7 Ω cm for the a-Si:H, the resistance of the quasi-neutral region is 10^3 Ω cm^2. Thus, 0.1 V will develop across this region when $J = 10^{-4}$ A/cm^2.

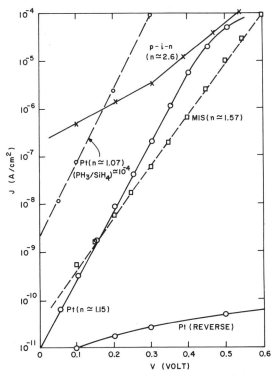

Fig. 6.6 Dark $I-V$ characteristics of a Pt Schottky-barrier diode, a MIS diode, and a $p-i-n$ diode. Also shown are data for a phosphorus-doped Pt Schottky-barrier diode.

The operating range of an a-Si:H diode can be extended by lightly doping the bulk of the film with phosphorus. For a discharge atmosphere with $PH_3/SiH_4 \simeq 10^{-4}$ the resistivity of the film is $\sim 10^3\ \Omega\,cm$, so the resistance of the quasi-neutral region ($\sim 1\ \mu m$ thick) is only 0.1 $\Omega\,cm$. In this case, the diode can accommodate a current density of 1 A/cm^2 before 0.1 V develops across the quasi-neutral region. As shown in Fig. 6.6, phosphorus doping causes an increase in the reverse saturation current density, but the diode quality is still good ($n \simeq 1.07$) [65].

When an a-Si:H diode is switched from forward to reverse bias, the recovery times are on the order of $\sim 10^{-1}$–10^{-4} s. Actually, the recovery cannot be characterized by a single time, because the distribution of trap states in the gap gives rise to a distribution of recovery times [66].

C. LIGHT-EMITTING DIODES

When an a-Si:H diode is forward-biased, electroluminescence can be detected [52]. The emission spectrum is identical to that obtained by photoluminescence, generally peaking at a photon energy of ~ 1.2–1.3 eV at 78 K. Figure 6.7 shows electroluminescence spectra for two currents through a Pt Schottky-barrier diode (area $\simeq 2\ mm^2$). The emission spectrum does not change as the current is varied but, as shown in Fig. 6.8, the emission intensity increases sublinearly with current [67]. At low currents, the external quantum efficiency was estimated to be 1.5×10^{-3} at 78 K [52]. Because of the high resistivity of undoped a-Si:H at 78 K, a forward bias of > 30 V was needed before electroluminescence was detected.

Although the performance of a-Si:H light-emitting diodes has not been characterized as a function of temperature, the behavior is probably similar to that of the photoluminescence. Thus, the emission intensity should fall with an activation energy of ~ 0.12 eV for temperatures above ~ 150 K. Small additions of phosphorus to the a-Si:H should reduce the

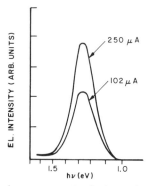

Fig. 6.7 Electroluminescence spectra for two values of the diode current.

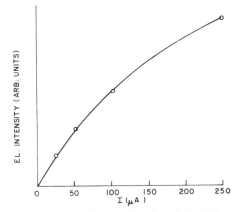

Fig. 6.8 Emission intensity as a function of the diode current.

activation energy [59] and, although the emission intensity should be reduced at 78 K by the doping, it may be larger than that produced by undoped films at room temperature.

3. Solar Cells

The increasing interest in amorphous semiconductors over the past few years has been largely generated by the promise of low-cost power from amorphous semiconductor solar cells. As mentioned in Section 2, the first amorphous semiconductor solar cell was fabricated in 1974, and the first published papers appeared in 1976 [68].

A solar cell is a semiconductor device that converts sunlight directly to electric power with a conversion efficiency η given by the expression

$$\eta = J_m V_m / P_i = (FF) J_{sc} V_{oc} / P_i \tag{6.7}$$

where J_m and V_m are the current density and voltage produced by a cell operating under maximum power conditions when illuminated by sunlight with a power density P_i, J_{sc} the short-circuit current density of the cell, and V_{oc} the open circuit voltage; FF is the fill factor and is defined by Eq. (6.7).

An amorphous semiconductor must exhibit several properties before it can be considered a candidate for a thin-film solar cell. The material must be capable of absorbing a significant fraction of the solar energy in the film thickness considered for the device. Moreover, the minority carrier diffusion length should be an appreciable fraction of the film thickness, or the depletion region should extend throughout most of the film. One must also be able to form a junction in the amorphous semiconductor as well as an ohmic contact. Finally, the conductivity of the material under illumination must be large enough to prevent a series resistance limitation in the quasi-neutral region.

To date a-Si:H is the only amorphous semiconductor that has exhibited all the above-mentioned properties. We will now consider the a-Si:H solar cell structures, their photovoltaic characteristics, and the stability of the devices.

A. STRUCTURES

Solar cells of a-Si:H have been fabricated in all the structures mentioned briefly in Section 4.2.a, but conversion efficiencies of 4–6% have only been attained in Schottky-barrier, MIS, and p–i–n cells. Generally, two types of substrates are used: metal sheet, such as stainless or cold-rolled steel, and glass coated with a metal film or a transparent conductive oxide such as indium–tin oxide (ITO).

A Schottky-barrier cell is constructed by first depositing a thin n^+ layer ($\simeq 0.1$ μm) of a-Si:H (from $\sim 0.1\%$ PH_3 in SiH_4) onto a metal or metalized glass substrate at $T_2 \simeq 200°–400°C$ (see Fig. 6.9). Then, an undoped layer of a-Si:H is deposited to a thickness of 0.3–2.0 μm. The Schottky barrier is formed by evaporating ~ 50 Å of a high-work-function metal such as Pt onto the undoped a-Si:H. Finally, the cell is coated with an antireflection layer such as ~ 450 Å of ZrO_2 [69]. For large-area cells, a current-collection grid is deposited onto the Pt before the antireflection coating. Collection grids have been formed by first depositing several hundred angstroms of Pt before depositing ~ 1 μm of Ag or Al. This additional Pt acts as a buffer layer to prevent shunting of the barrier by the Ag or Al.

Schottky-barrier cells with open-circuit voltages in excess of 300 mV have been fabricated with metals such as Pt, Pd, Ir, Rh, Re, and Au. Dopants such as P, As, Sb, and Cs have been used to form an n^+ layer, and materials such as ITO, SnO_2 (Sb), Si_3N_4, and ZrO_2 have been used as

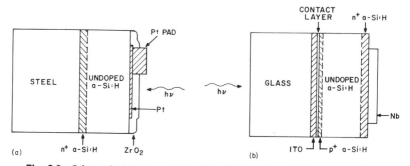

Fig. 6.9 Schematic drawings of (a) a Schottky-barrier and (b) a p–i–n cell.

antireflection coatings. Back-illuminated Schottky-barrier cells have been fabricated by using substrates of semitransparent metal films on glass. However, the efficiency of this type of cell was limited to $\sim 1.0\%$ because most of the holes generated near the back contact did not reach the Schottky barrier on the other side of the film [69]. Schottky-barrier cells have also been fabricated by evaporating semitransparent films of Al, Cr, or Ti on p-type a-Si:H with a thin p^+ layer on a Pt or Pd substrate. These cells also exhibited relatively low values of J_{sc} (< 2 mA/cm^2 without antireflection coatings).

Recently, another type of Schottky-barrier cell has been fabricated by sputter-depositing ~ 100 Å of a Pt–SiO$_2$ cermet onto undoped a-Si:H [20]. The cermet has a high resistivity ($\sim 10^2$–10^4 Ω cm) and is overcoated with a conductive transparent oxide such as ITO to act both as a current-collection layer and an antireflection coating.

An MIS cell is similar to a Schottky-barrier cell except that a thin insulating layer is located between the metal and the semiconductor. The insulating layer is typically on the order of 20–30 Å thick, so that photogenerated carriers are able to tunnel through it. Many of the earlier Schottky-type cells were actually MIS cells, since they were exposed to air before the Pt was evaporated onto the a-Si:H. Etching the a-Si:H in buffered HF just prior to the Pt evaporation allows the formation of a reasonably intimate Schottky-barrier contact. An efficient MIS cell can be fabricated by heating the a-Si:H film for 15 min at 350°C in air before depositing the Pt. MIS cells have also been fabricated with insulating layers of Si$_4$N$_3$ [70] and TiO$_2$ [71]; in the latter case, the semitransparent metal was a thin film of Ni.

The earliest p–i–n cells were fabricated on glass substrates coated with ITO [68]. However, the contact resistance between the ITO and the p^+ layer is often large, perhaps because of the formation of a thin resistive oxide layer. This resistance can be reduced by the deposition of a thin, semitransparent metal film [69] or a cermet [20] on top of the ITO. The p^+ layer is typically a few hundred angstroms thick and is deposited from a discharge containing ~ 0.1–1.0% of B$_2$H$_6$ in SiH$_4$. The thickness of the undoped layer is usually in the range of 0.3–2.0 μm, while that of the n^+ layer is typically 0.1 μm. All the a-Si:H layers can be deposited at substrate temperatures in the range of 200°–400°C (~ 300°C is usually optimum). The top contact is normally an evaporated film of Nb, Al, Ti, or Mo.

We have also fabricated p–i–n cells in a manner similar to that of the Schottky-barrier cells, except that a thin p^+ layer is deposited on top of the undoped a-Si:H. One can also fabricate a p–i–n cell so that the light is incident on the n^+ side of the cell by using a semitransparent top electrode.

B. PHOTOVOLTAIC CHARACTERISTICS

A solar cell can be easily characterized by measuring the $I-V$ dependence under illumination comparable to 1 sun (AM1, ~ 100 mW/cm^2). A series of illuminated $I-V$ curves is shown in Fig. 6.10 for Ir Schottky-barrier cells made in a dc discharge with a cathodic screen (see Section 6.3.1). The $I-V$ characteristics show an improvement as the screen–substrate spacing decreases from 3.0 to 0.5 cm. The hydrogen content varies only from 12 to 10 at.% over the same range of spacings. The decrease in the photovoltaic parameters with increasing spacing is probably due to an increase in polymerization in the discharge.

The conversion efficiency of the cell represented by curve a in Fig. 6.10 is $\sim 6.0\%$ in simulated sunlight of 100 mW/cm^2. The short-circuit current density of 14.5 mA/cm^2 is the largest observed to date. The large value of V_{oc} (753 mV) indicates that the device is actually an MIS device because of a thin nascent oxide layer beneath the Ir.

The largest value of V_{oc} observed to date is ~ 895 mV for both an MIS device and a $p-i-n$ device, but in both cases the fill factors were poor (< 0.4). The best fill factor (0.674) was observed in a Pt Schottky-barrier cell for AM1 illumination, but values of ~ 0.7 have been observed using blue light. If the best values of J_{sc}, V_{oc}, and FF could be obtained in the same cell, then the conversion efficiency would be $\sim 8.8\%$.

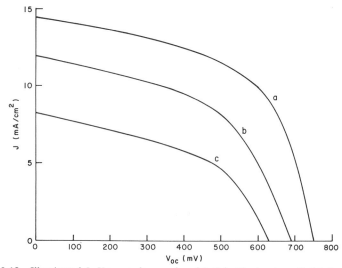

Fig. 6.10 Illuminated $I-V$ curves for a series of Ir Schottky-barrier cells fabricated in a dc discharge where the spacing between the cathodic screen and the substrate decreases from 3.0 cm for curve c to 0.5 cm for curve a; the antireflection coating was ~ 450 Å of ZrO$_2$.

The theoretical limit for the conversion efficiency of an a-Si:H solar cell can be roughly determined by estimating the limits for J_{sc}, V_{oc}, and *FF*. From the optical absorption data (see Fig. 6.3), one can calculate that the maximum short-circuit current density for AM1 illumination should be on the order of 22 mA/cm² for an a-Si:H film ~ 1.5 μm thick [68]. This calculation assumes that every photon absorbed in the film creates an electron–hole pair that is collected by the junction. Since the optical absorption increases as the hydrogen content decreases, the estimate of the maximum value of J_{sc} will vary as the hydrogen content varies.

The maximum value of V_{oc} that might be expected in an a-Si:H solar cell is more difficult to estimate, since the density and distribution of gap states are not well known. However, the doping studies of Spear and LeComber [51] and the field-effect measurements of Madan *et al.* [46] indicate that tail states might limit the maximum value of the built-in potential to ~ 1.2 eV. This value is close to the maximum built-in potential observed in *p–i–n* devices [72] and also to the maximum observed value of the Schottky-barrier height [64]. This limitation on the built-in potential indicates that the maximum observed value of $V_{oc} \approx 900$ mV may be close to the practical limit.

The theoretical limit for the fill factor is ~ 0.87, assuming an ideal diode [73]. Thus, for incident radiation of 100 mW/cm², Eq. (6.7) predicts $\eta \simeq 0.87(22)(0.9)/0.1 \simeq 17\%$, assuming no reflection losses and no absorption losses in the transparent, current-collecting electrode.

The observed conversion efficiencies are limited to < 6% mainly as a result of poor minority carrier diffusion lengths. The active region or collection region W of an a-Si:H solar cell can be estimated from the following expression for the short-circuit photocurrent generated by light of wavelength λ:

$$J_{sc}(\lambda) = \phi(\lambda)\left\{1 - \exp\left[-\alpha(\lambda)W\right]\right\} \qquad (6.8)$$

where $\phi(\lambda)$ is the photon flux entering the a-Si:H and $\alpha(\lambda)$ the absorption coefficient. Figure 6.11 shows a fit of Eq. (6.8) to the collection efficiency of a Pt Schottky-barrier solar cell (~ 100 Å of Pt; no antireflection coating) assuming a collection region of ~ 0.3 μm [42]. The collection efficiency is defined as the percentage of electron–hole pairs collected by the junction per incident photon. Capacitance–voltage measurements indicate that the depletion width is on the order of ~ 0.2 μm, so that the minority carrier diffusion length is ~ 0.1 μm. Thus, the space charge region plays an important role in collecting the minority carriers in a-Si:H solar cells. In general, the depletion region decreases as the light intensity increases, as a result of the trapping of photogenerated holes [67].

An antireflection coating increases the collection efficiency but com-

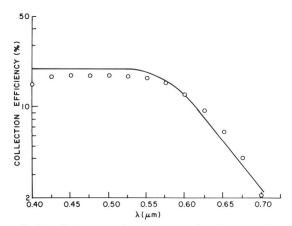

Fig. 6.11 Collection efficiency as a function of wavelength for a Pt Schottky-barrier cell. Circles, measured collection efficiency; solid line, calculated maximum efficiency for 0.3 μm of a-Si:H (normalized to 20% transmission).

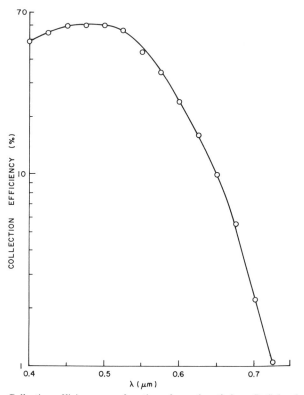

Fig. 6.12 Collection efficiency as a function of wavelength for a Pt Schottky-barrier cell with a ZrO_2 antireflection coating.

plicates the analysis described in the previous paragraph, since the transmission (or reflection) is wavelength-dependent. Figure 6.12 shows the collection efficiency of a Pt Schottky-barrier cell with a ZrO_2 antireflection coating. The measured short-circuit current density for this cell was 10 mA/cm², and a similar value is obtained by integrating the collection efficiency response in Fig. 6.12 over the AM1 solar spectrum.

The quality of an illuminated a-Si:H junction can be determined by plotting the short-circuit current as a function of the open-circuit voltage; V_{oc} is varied by changing the light intensity. Figure 6.13 shows representative $J_{sc}-V_{oc}$ plots for a Pt Schottky-barrier cell, a MIS cell, and a $p-i-n$ device. In all cases, the diode quality factor for the illuminated junction n' is closer to unity than the diode quality factor determined from the dark $I-V$ characteristics (see Fig. 6.6). The diode quality factor for the MIS cell changes from 1.50 to 1.04 at $V_{oc} \simeq 0.49$ V, possibly because of the saturation of interface states.

The photovoltaic characteristics depend strongly on the a-Si:H deposition conditions. The short-circuit current density is very small for substrate

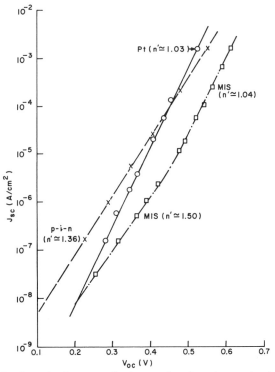

Fig. 6.13 The short-circuit current density as a function of open-circuit voltage for a Pt Schottky-barrier cell, a MIS cell, and a $p-i-n$ cell.

TABLE 6.1

V_{oc} and J_{sc} as a Function of RF Power[a]

Relative power	V_{oc} (mV)	J_{sc} (mA/cm^2)
100	652	5.25
180	748	4.5
260	775	4.0

[a] $T_s \simeq 320°C$; SiH$_4$ pressure $\simeq 0.8$ torr.

temperatures T_s less than $\sim 200°C$, shows a gradual increase as T_s increases from ~ 200 to $\sim 400°C$, and then falls rapidly for $T_s > 400°C$ [53]. This behavior can be related to the defects described in Section 6.3.3. The photovoltaic characteristics also vary with deposition parameters such as discharge power, SiH$_4$ pressure, flow rate, substrate bias, and doping level. Table I shows how V_{oc} and J_{sc} for a series of Pt Schottky-barrier cells varied as the power of an rf electrodeless discharge was changed. The relative power settings are the same as described in Section 6.3.2 for Fig. 6.2. The variations in V_{oc} and J_{sc} appear to be caused mainly by changes in the hydrogen content of the films. As the hydrogen content increases, the optical gap increases, causing J_{sc} to decrease and V_{oc} to increase [34].

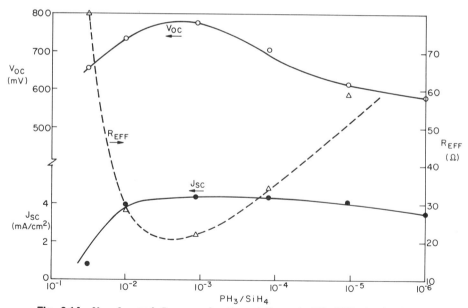

Fig. 6.14 V_{oc}, J_{sc}, and R_{eff} as a function of the ratio PH$_3$/SiH$_4$ in the discharge atmosphere during the deposition of the n^+-contacting layer. Here R_{eff} is an effective series resistance determined from the slope at $I = 0$.

The performance of an a-Si:H solar cell is strongly affected by the doping levels in doped regions of the cell. Figure 6.14 shows the variation of some photovoltaic parameters with the doping level of phosphorus in the n^+-contacting layer of a Pt Schottky-barrier cell. The n^+ layer was ~ 180 Å thick and was deposited in a dc discharge at $T_s \simeq 320°C$. The optimum doping level ($PH_3/SiH_4 \simeq 2 \times 10^{-3}$) is close to that determined by Spear and LeComber [54] for obtaining the maximum conductivity in phosphorus-doped a-Si:H.

In a p–i–n cell, the doping level in the p-layer strongly influences both V_{oc} and J_{sc}. A large value of V_{oc} is obtained if the p-layer is thick and heavily doped with boron, but J_{sc} will be small because of recombination in the p-layer [72]. Even relatively small amounts of phosphorus or boron (~ 100 ppm) in the "undoped" bulk of an a-Si:H cell can reduce J_{sc}, apparently as a result of increased recombination of photogenerated carriers. However, a-Si:H solar cells are relatively insensitive to impurities such as carbon and nitrogen [53].

The larger photoconductive effect in a-Si:H (see Section 3.3) allows an a-Si:H solar cell to operate at temperatures as low as ~ 200 K without any serious degradation in efficiency. At lower temperatures, the increasing resistance of the quasi-neutral region causes both J_{sc} and the fill factor to decrease [53].

C. STABILITY

A low-cost solar cell must be a stable electronic device with an operational lifetime of at least 20 years. Since a-Si:H is a relatively new material and a-Si:H solar cells even newer, we now address the important issue of stability in the light of our present understanding.

As mentioned in Section 6.3.2, hydrogen diffusion can cause degradation at elevated temperatures in short times (< 1 h at 400°C) because of the creation of dangling bonds. However, the activation energy of ~ 1.5 eV indicates that similar degradation will not occur at temperatures less than 100°C until after more than 10^4 years [30].

Degradation due to the diffusion of dopants does not appear to be a potential problem. Recent compositional profiles performed at RCA Laboratories indicate that the diffusion coefficients of boron and phosphorus are $< 3 \times 10^{-17}$ cm^2/s at 400°C in a-Si:H. Moreover, the diffusion coefficients of many metals, such as Mo, Nb, Ta, and Cr, are $< 10^{-18}$ cm^2/s at 400°C. The value for Fe is also $< 10^{-18}$ cm^2/s at 300°C but increases to $\sim 10^{-15}$ cm^2/s at 400°C. Thus, Fe is a suitable substrate for depositions at 300°C but not 400°C. Aluminum exhibits significant diffusion even at 300°C and is generally not a good substrate. Palladium exhibits a diffusion coefficient of $\sim 10^{-15}$ cm^2/s at 180°C [67]. Thus, one must be careful in choosing electrode materials to avoid potential long-term problems due to diffusion.

The diffusion coefficient of oxygen has been estimated to be $\sim 6 \times 10^{-18}$ cm^2/s at 450°C, but a thin nascent oxide (~ 10 Å) forms upon exposure of the a-Si:H to air. Since the a-Si:H contains large quantities of hydrogen, this oxide layer probably contains a large concentration of OH groups.

Generally, we find that MIS devices that use the nascent oxide as a thin insulating layer exhibit degradation upon exposure to humidity [69]. Some Schottky-barrier devices are also sensitive to humidity [53]. This degradation can be prevented by storing the devices in an inert atmosphere or by encapsulation.

4. Photodetectors and Photometers

Diodes of a-Si:H can also be tailored to act as photodetectors or photometers; both devices are similar to solar cells. In the case of a photodetector, one may desire a high-frequency response. The response

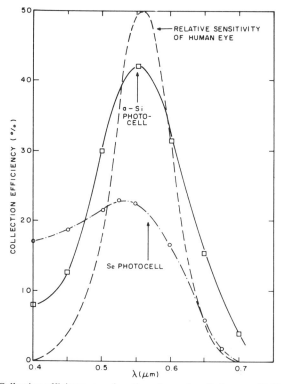

Fig. 6.15 Collection efficiency as a function of wavelength for an a-Si:H $p–i–n$ cell and a Se photocell. The relative wavelength sensitivity of the human eye is also shown.

time depends on the drift mobility of photogenerated carriers μ, the film thickness d, and the applied voltage V through the relationship

$$\tau_{\text{response}} = d^2/\mu V \qquad (6.9)$$

For an a-Si:H film 0.1 μm thick, $\tau_{\text{response}} \simeq 10^{-7}$ s for an applied voltage of 1 V and a hole drift mobility of $\sim 10^{-3}$ cm^2/V s [45]. Thus, this type of device should be sensitive to frequencies of ~ 10 MHz. The spectral response will be similar to that of a solar cell (see Section 6.4.3), but with less sensitivity at long wavelengths if the photodetector film is thin.

A photometer is a photovoltaic device that measures brightness; i.e., the spectral response of the device is similar to the sensitivity of the human eye. The spectral response of a p–i–n cell can be tailored to that of the human eye by adjusting the thickness of the p- and i-layers and the doping level in the p-layer. Figure 6.15 shows the collection efficiency of a p–i–n device with a thicker p-layer than that used in solar cells.

The collection efficiency has a wavelength dependence similar to that of the human eye, and the device could be tailored to provide a much closer fit. For comparison, the collection efficiency of a Se photovoltaic device is also shown in Fig. 6.15. This type of device is commonly used in light meters.

5. Optical Storage in a-Si:H

The reversible, light-induced conductivity changes in a-Si:H (see Section 6.3.3) can be used to store information. The dark conductivity of an annealed film of a-Si:H can be decreased locally by exposure to light [38]. An even more intense light source can be used to heat the film locally and restore the original conductivity. However, long exposures are required to produce a significant change in conductivity; an exposure of ~ 15 min of white light of ~ 200 mW/cm^2 is needed to decrease the dark conductivity by an order of magnitude.

Films of a-Si:H deposited at low substrate temperatures ($\sim 50°$C) exhibit an irreversible change in optical properties when exposed to an intense laser beam [74]. The absorption edge shifts to lower energies because of the local heating of the film and the subsequent evolution of hydrogen. Films deposited at low substrate temperatures may be more accurately classified as amorphous silicon subhydride than as amorphous silicon [17]. The change in composition due to laser beam heating causes a change in the etch rates of the exposed and unexposed portions of a film. Thus, a-Si:H films deposited at low substrate temperatures can act as either permanent optical storage mediums or as a photoresist material.

6. Thin-Film Transistors

As mentioned in Section 6.2, Neudeck and Malhotra [14] have fabricated TFTs using evaporated amorphous silicon. However, these devices required large gate voltages (> 50 V) to produce significant modulation in the drain current. While they did not report on the frequency response of their TFTs, it is likely that the devices were very slow as a consequence of the large density of defects in the evaporated amorphous silicon.

Spear and LeComber [75] used a similar field-effect type of device to determine the density of states in a-Si:H. However, in that early work, a glass substrate (~ 250 μm thick) was used as the insulating layer between the gate electrode and the a-Si:H. Thus gate voltages as large as 1 kV had to be applied to modulate the drain current by a factor of 10^3. In later work by Madan et al. [46] a thin film of amorphous Si_3N_4 (~ 1–3 μm thick) was used as the insulating layer, and then gate voltages of ~ 50 V were able to produce a modulation of $\sim 10^3$ in the drain current.

More recently, L. A. Goodman [76] at RCA Laboratories has succeeded in fabricating a-Si:H TFTs using a metal-oxide semiconductor (MOS) structure (see Fig. 6.16). The source and drain contacts were made by depositing a thin layer (~ 300 Å) of phosphorus-doped a-Si:H on top of a Cr film on glass and then using photolithography to define the electrode pattern. The spacing between the source and drain was 10 μm, and the channel width was 320 μm. The undoped a-Si:H was usually deposited to a thickness in the range of 0.1–0.5 μm. The SiO_2 layer was typically 0.1–0.2 μm thick and was deposited either by electron-beam evaporation or by sputter deposition. The Cr gate electrode was also defined using photolithography.

Figure 6.17 shows a plot of the drain I–V characteristics for various values of the gate voltage V_G for a TFT constructed with 0.1 μm of undoped a-Si:H ($T_s \simeq 375°C$) and 0.2 μm of sputtered SiO_2. The threshold voltage V_T for this type of TFT was typically in the range of 3–8 V. Good saturation was observed when the drain voltage approached $V_G - V_T$. The drain current could be modulated by a factor of 10^5–10^6 by varying the gate voltage from -5 to $+40$ V. The transistor characteristics indicated that the extended-state electron mobility was > 1 cm^2/V s [76].

The time response of such a transistor when a voltage pulse is applied across the source and drain is shown in Fig. 6.18. Most of the switching occurs within a time of ~ 10 μs, but a tail can always be observed that

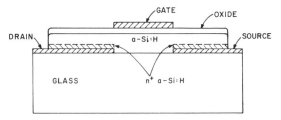

Fig. 6.16   Schematic drawing of an a-Si:H MOS TFT.

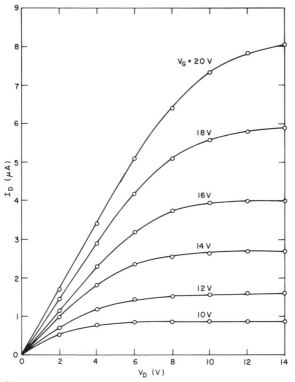

Fig. 6.17 The drain current as a function of the drain voltage for various values of the gate voltage.

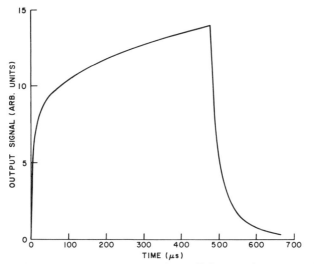

Fig. 6.18 The output signal for a 30-V pulse applied across the source and drain. The gate voltage was 30 V, and the load resistance was 10^5 Ω.

extends out to times on the order of milliseconds. This tail is apparently due to a distribution of trapping states in the a-Si:H.

A thin-film metal-base transistor using a-Si:H has been fabricated by Deneuville and Brodsky [77]. This structure consists of a thin metal film (~ 100 Å of Pt) sandwiched between two undoped a-Si:H layers; ohmic contacts were made to the undoped layers by depositing n^+ a-Si:H at the beginning and end of the fabrication sequence. This type of transistor actually consists of two Schottky diodes back to back, and one diode is used as the emitter and the other as the collector. The observed transistor characteristics were poor, and the injection ratio (variation of excess collector current to emitter current) was limited to $< 8\%$. The poor performance was attributed mainly to the poor characteristics of the top diode.

7. Future Directions

All the a-Si:H devices we have discussed are still in the experimental stage. The a-Si:H solar cell has the greatest potential because of its promise of low manufacturing costs and the anticipated demand for new sources of energy in the future. At this time the stability of a-Si:H solar cells does not appear to be a problem, so improving the conversion efficiency will be the most important objective for the next few years.

The quality of a-Si:H will undoubtedly be improved by careful optimization of the discharge conditions. There is also the possibility that the addition of certain dopants or impurities might improve the minority carrier transport in the material. Other techniques may be developed for making good-quality a-Si:H, e.g., hydrogenation of sputtered or evaporated amorphous silicon by exposure to a hydrogen discharge [78].

At the present time a-Si:H is the only amorphous semiconductor that has displayed good photovoltaic characteristics. Although a-Ge:H is somewhat similar to a-Si:H, a-Ge:H appears to possess a significantly larger density of gap states [79]. Other types of amorphous semiconductors are being investigated such as hydrogenated amorphous SiC [80] and hydrogenated amorphous arsenic [18]. Moreover, considerable research is being conducted in the area of amorphous chalcogenide semiconductors, and a better understanding of these materials is being developed [81].

Many amorphous semiconductors are fabricated under highly nonequilibrium conditions (e.g., glow-discharge deposition) that allow the incorporation of elements in quantities not possible in the crystalline state. Thus, the composition and properties of amorphous semiconductors can be varied over a wide range, and future efforts should result in the development of new amorphous semiconductor materials with useful properties.

As mentioned earlier in the case of solar cells, amorphous semiconductors may offer a distinct cost advantage in any application that requires large areas of semiconductor films. Amorphous semiconductor TFTs may eventually find applications in large-area displays or flat-panel televisions. Another possibility is that relatively efficient electroluminescent devices might be developed for display applications; visible photoluminescence has recently been observed in hydrogenated amorphous SiC at 77 K [82]. Other potential applications for amorphous semiconductors include thermoelectric energy conversion and photothermal energy conversion.

The present challenge for scientists working in the area of amorphous semiconductors is to attain a better understanding of these materials so that the experimental devices discussed in this article will become a practical reality. Moreover, an improved understanding of amorphous semiconductors should allow new devices and applications to be developed in the future.

REFERENCES

1. N. F. Mott and E. A. Davis, "Electronic Processes in Non-Crystalline Materials." Oxford Univ. Press (Claredon), London and New York, 1971.
2. J. H. Dessauer and H. E. Clark, "Xerography and Related Processes." Focal Press, New York, 1975.
3. N. Goto, Y. Isozaki, K. Shidara, E. Maruyama, T. Hirai, and T. Fujita, *IEEE Trans. Electron. Devices* **ED-21**, 662 (1974).
4. L. Young, "Anodic Oxide Films," p. 147. Academic Press, New York, 1961.
5. S. R. Ovshinsky, *J. Non-Cryst. Solids* **2**, 99 (1970).
6. S. H. Holmberg and M. P. Shaw, *Proc. Int. Conf. Amorphous Liquid Semicond. 5th* p. 687. Taylor and Francis, London, 1974.
7. D. D. Thornberg, R. I. Johnson, and T. M. Hayes, *Bull. Am. Phys. Soc.* **23**, 681 (1978).
8. R. B. South, J. M. Robertson, and A. E. Owen, *Proc. Int. Conf. Amorphous Liquid Semicond., 7th* p. 722. CICL, Edinburgh, 1977.
9. S. R. Ovshinsky, *Phys. Rev. Lett.* **21**, 1450 (1968).
10. J. S. Berkes, S. W. Ing, Jr., and W. Hillegas, *J. Appl. Phys.* **42**, 4908 (1971).
11. J. P. de Neufville, *Proc. Int. Conf. Amorphous Liquid Semicond. 5th* p. 1351. Taylor and Francis, London, 1974.
12. J. Hajto, J. Gazso, and G. Zentai, *Proc. Int. Conf. Amorphous Liquid Semicond., 7th* p. 807. CICL, Edinburgh, 1977.
13. L. Mei and J. E. Greene, *J. Vac. Sci. Technol.* **11**, 145 (1974).
14. G. W. Neudeck and A. K. Malhotra, *Solid-State Electron.* **19**, 721 (1976).
15. R. C. Chittick, J. H. Alexander, and H. F. Sterling, *J. Electrochem. Soc.* **116**, 77 (1969).
16. D. E. Carlson, U.S. Patent No. 4064521 (1977).
17. M. H. Brodsky, *Thin Solid Films* **40**, L23 (1977).
18. J. C. Knights, *Bull. Am. Phys. Soc.* **23**, 295 (1978).
19. J. C. Knights, *Phil. Mag.* **34**, 663 (1976).
20. J. J. Hanak and V. Korsun, *Proc. IEEE Photovoltaics Spec. Conf., 13th* p. 780. IEEE, New York, 1978.

21. G. A. N. Connell and J. R. Pawlik, *Phys. Rev. B* **13**, 787 (1976).
22. A. K. Malhotra and G. W. Neudeck, *Appl. Phys. Lett.* **28**, 47 (1976).
23. J. I. Pankove, Private communication (1978).
24. A. J. Lewis, G. A. N. Connell, W. Paul, J. R. Pawlik, and R. J. Temkin, *Proc. Int. Conf. Tetrahedrally Bonded Amorphous Semicond.* p. 27. American Institute of Physics, New York, 1974.
25. T. Sakurai and H. D. Hagstrum, *Phys. Rev. B* **12**, 5349 (1975).
26. A. Triska, D. Dennison, and H. Fritzche, *Bull. Am. Phys. Soc.* **20**, 392 (1975).
27. M. H. Brodsky, M. Cardova, and J. J. Cuomo, *Phys. Rev. B* **16**, 3556 (1977).
28. P. J. Zanzucchi, C. R. Wronski, and D. E. Carlson, *J. Appl. Phys.* **48**, 5227 (1977).
29. H. Fritzche, D. D. Tsai, and P. Persans, *Solid-State Technol.* **21**, 55 (1978).
30. D. E. Carlson and C. W. Magee, *Appl. Phys. Lett.* **33**, 81 (1978).
31. A. Barna, P. B. Barna, G. Rodroczi, L. Toth, and P. Thomas, *Phys. Status Solidi (a)* **41**, 81 (1977).
32. M. H. Tanielian, J. Fritzsche, and C. C. Tsai, *Bull. Am. Phys. Soc.* **22**, 336 (1977).
33. V. A. Singh, C. Weigel, J. W. Corbett, and L. M. Roth, *Phys. Status Solidi (b)* **81**, 637 (1977).
34. D. E. Carlson, C. W. Magee, and A. R. Triano, *J. Electrochem. Soc.* **126**, 688 (1979).
35. B. von Roedern, L. Ley, and M. Cardona, *Phys. Rev. Lett.* **39**, 1956 (1977).
36. E. A. Davis, "Amorphous Semiconductors" (P. G. LeComber and J. Mort, eds.), p. 450. Academic Press, New York, 1973.
37. R. J. Loveland, W. E. Spear, and A. Al-Sharloty, *J. Non-Cryst. Solids* **13**, 55 (1973/1974).
38. D. L. Staebler and C. R. Wronski, *Appl. Phys. Lett.* **31**, 292 (1977).
39. P. G. LeComber, D. I. Jones, and W. E. Spear, *Phil. Mag.* **35**, 1173 (1977).
40. W. Rehm, R. Fisher, J. Stuke, and H. Wagner, *Phys. Status Solidi (b)* **79**, 539 (1977).
41. J. G. Simmons, *Phys. Rev.* **155**, 657 (1967).
42. C. R. Wronski and D. E. Carlson, private communication (1978).
43. A. Rose, "Concepts in Photoconductivity and Allied Problems." Wiley (Interscience), New York, 1963.
44. C. R. Wronski and D. E. Carlson, *Proc. Int. Conf. Amorphous Liquid Semicond., 7th,* p. 452. CICL, Edinburgh, 1977.
45. A. R. Moore, *Appl. Phys. Lett.* **31**, 762 (1977).
46. A. Madan, P. G. LeComber, and W. E. Spear, *J. Non-Cryst. Solids* **20**, 239 (1976).
47. J. C. Knights and D. K. Biegelson, *Solid State Commun.* **22**, 133 (1977).
48. C. R. Wronski and D. E. Carlson, *Solid State Commun.* **23**, 421 (1977).
49. C. R. Wronski, *IEEE Trans. Electron Dev.* **ED-24**, 351 (1977).
50. J. J. Hanak, P. J. Zanzucchi, D. E. Carlson, C. R. Wronski, and J. I. Pankove, *Proc. Int. Vac. Congr., 7th*; *Int. Conf. Solid Surfaces, 3rd, Vienna* (1977).
51. W. E. Spear and P. G. LeComber, *Solid State Commun.* **17**, 1193 (1975).
52. J. I. Pankove and D. E. Carlson, *Appl. Phys. Lett.* **29**, 620 (1976).
53. D. E. Carlson, *Tech. Digest IEEE Int. Electron Dev. Meeting* p. 214. IEEE, New York, 1977.
54. W. E. Spear and P. G. LeComber, *Phil. Mag.* **33**, 935 (1976).
55. W. E. Spear, *Adv. Phys.* **26**, 811 (1977).
56. P. G. LeComber, D. I. Jones, and W. E. Spear, *Phil. Mag.* **35**, 1173 (1977).
57. D. I. Jones, P. G. LeComber, and W. E. Spear, *Phil. Mag.* **36**, 541 (1977).
58. D. Engemann and R. Fisher, "Structure and Excitations in Amorphous Solids" (G. Lucovsky and F. L. Galeener, eds.), p. 37. American Institute of Physics, New York, 1976.

59. T. S. Nashashibi, I. G. Austin, and T. M. Searle, *Phil. Mag.* **35**, 831 (1977).
60. W. Beyer and H. Mell, *Proc. Int. Conf. Amorphous Liquid Semicond., 7th* p. 333. CICL, Edinburgh, 1977.
61. M. H. Tanielian, E. M. D. Symbalisty, and H. Fritzsche, *Bull. Am. Phys. Soc.* **22**, 336 (1977).
62. S. J. Fonash, *J. Appl. Phys.* **47**, 3597 (1976).
63. S. M. Sze, "Physics of Semiconductor Devices." Wiley (Interscience), New York, 1969.
64. C. R. Wronski, D. E. Carlson, and R. E. Daniel, *Appl. Phys. Lett.* **29**, 602 (1976).
65. R. W. Smith, private communication (1978).
66. R. S. Crandall, to be published in *J. Electron. Mater.* (1980).
67. D. E. Carlson, C. R. Wronski, J. I. Pankove, P. J. Zanzucchi, and D. L. Staebler, *RCA Rev.* **38**, 211 (1977).
68. D. E. Carlson and C. R. Wronski, *Appl. Phys. Lett.* **28**, 671 (1976).
69. D. E. Carlson, *IEEE Trans. Electron Dev.* **ED-24**, 449 (1977).
70. W. A. Anderson, J. K. Kim, S. L. Hyland, and J. Coleman, *Proc. IEEE Photovoltaics Spec. Conf., 13th*, p. 755. IEEE, New York, 1978.
71. J. I. B. Wilson, J. McGill, and S. Kimmond, *Nature (London)* **272**, 153 (1978).
72. D. E. Carlson and C. R. Wronski, *J. Electron. Mater.* **6**, 95 (1977).
73. J. Lindmeyer, COMSAT *Tech. Rev.* **2**, 105 (1972).
74. D. L. Staebler, *J. Appl. Phys.* **50**, 3648 (1979).
75. W. E. Spear and P. G. LeComber, *J. Non-Cryst. Solids* **11**, 219 (1972).
76. L. A. Goodman, private communication (1977).
77. A. Deneuville and M. H. Brodsky, *Bull. Am. Phys. Soc.* **23**, 247 (1978).
78. J. I. Pankove, M. A. Lampert, and M. L. Tarng, *Appl. Phys. Lett.* **32**, 439 (1978).
79. D. I. Jones, W. E. Spear, and P. G. LeComber, *J. Non-Cryst. Solids* **20**, 259 (1976).
80. D. A. Anderson and W. E. Spear, *Phil. Mag.* **35**, 1 (1977).
81. S. R. Ovshinsky and D. Alder, *Contemp. Phys.* **19**, 109 (1978).
82. D. Engemann, R. Fischer, and J. Knecht, *Appl. Phys. Lett.* **32**, 567 (1978).

7 | Industrial Applications of Polycrystalline Thin-Film Devices

R. A. MICKELSEN

The Boeing Aerospace Company
Seattle, Washington

7.1 INTRODUCTION

Devices based upon polycrystalline semiconductor thin films have existed for a long period of time. For the most part, they have been more in the nature of experimental or research components with limited commercial application. Thin-film materials, film preparation, and device design technology have, however, progressed to the point where these devices are commanding increasing attention and consideration by the electronics industry. There can be little doubt regarding the promising

209

future of many thin-film device concepts and their eventual widespread usage. Thin-film solar cells (as discussed in Chapter 8), active devices, and several optical instruments are prime examples relating to immediate or reasonably near-term applications of substantial industrial importance.

The purpose of this chapter is to provide a summarized review of the polycrystalline semiconductor thin-film materials and devices which have been extensively developed and have advanced to the technology forefront. For these selected cases, a description will be given of the most useful materials, pertinent material properties, applied film preparation methods, device configurations, and significant device performance characteristics.

7.2 TRANSISTORS

The thin-film transistor (TFT) is one of the most highly developed and commercially promising applications for deposited polycrystalline semiconductor films. Originally intended as an active device to be incorporated with thin-film passive components in the fabrication of complex integrated circuits, the rapid development of single-crystal silicon metal-oxide semiconductor field-effect transistor (MOS FET) technology resulted in diminished interest in the device. However, this trend has recently been reversed as a result of efforts to build large, active matrices for display and sensor systems. The ability to form complex circuits containing thousands of active and passive components on a large-area glass substrate entirely with thin-film processing methods is a particularly attractive approach from the viewpoint of performance and manufacturing cost.

Of the several types of TFTs investigated to date, the field-effect device developed by Weimer [1, 2] has shown the greatest potential for becoming an acceptable semiconductor device. Many different versions of the field-effect transistor have been studied, but these differ mainly in electrode positioning and sequencing of the film layers. A common structure (staggered electrode) is shown in Fig. 7.1 and is sufficient to depict the major features for device operation.

Typically, the source-drain electrodes are 0.1–0.2 cm in length and separated by a gap in the 5–10 μm range. Film thicknesses for the semiconductor and insulator are 0.15–1.0 μm and 100–2000 Å, respectively. By applying voltage to the gate electrode, the carrier concentration in a thin channel of the semiconductor beneath the gate insulator is enhanced or depleted depending upon the voltage polarity and semiconductor conductivity type. Consequently, it is possible to modulate the source-drain current effectively over many orders of magnitude by the application of a small gate voltage. An expression for the effect [2] is

$$g_m = \partial I_d / \partial V_g = \mu C_g V_d / L^2 \qquad (7.1)$$

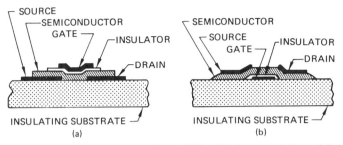

Fig. 7.1   Cross-sectional diagrams of two TFTs with "staggered-electrode" configurations. (From Weimer [2].) (a) Source–Drain electrodes deposited first; (b) Gate electrode deposited first.

where g_m is the transconductance of the device—an important figure of merit, $\partial I_d / \partial V_g$ the change in source-drain current flow per unit change in gate voltage, μ the effective drift mobility of the semiconductor, C_g the gate capacitance, V_d the source-drain voltage, and L the source-drain separation.

The drain characteristics of the TFT have been categorized by Weimer [2, 3] into two basic types—depletion and enhancement. In a depletion-type device, the semiconductor film displays a high conductivity which can be reduced by the application of gate voltage. In contrast, an enhancement mode device initially has very low conductance, but this can be greatly increased with gate voltage. Examples of the drain characteristics of these two devices for an *n*-type semiconductor film are presented in Fig. 7.2. A *p*-type film exhibits similar characteristics, but in the third instead of the first quadrant.

Saturation of the drain current at high drain voltages apparent in the curves is caused by a pinch-off of the conducting channel in the vicinity of the drain electrode. Weimer [2, 4] has described this effect by a parameter

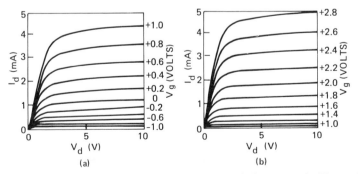

Fig. 7.2   Drain characteristics of CdS TFTs. (a) Depletion-type unit ($V_0 \sim -1$ V); (b) enhancement-type unit ($V_0 \sim +1$ V). (From Weimer [3].)

V_0, designated the threshold gate voltage required for the onset or pinch-off of drain current. In terms of the semiconductor and device parameters V_0 is expressed as

$$V_0 = - N_0 q / C_g \qquad (7.2)$$

where q is the electronic charge and C_g the gate capacitance; N_0 (for a depletion device) is the total number of free carriers in the gap region of the semiconductor at zero gate bias or (for an enhancement unit) represents the total number of traps which must be filled prior to achieving an increase in free-carrier density with an increase in gate voltage. The sign of N_0 is negative (V_0 positive) for enhancement devices, and the reverse for depletion-type units. While the magnitude of V_0 can be several volts positive or negative, normal procedure is to use processing techniques and device designs which make the V_0 value near zero.

1. Thin-Film Transistor Fabrication

By far, the predominant method for the fabrication of TFT devices involves vacuum evaporation and deposition onto glass substrates. Both evaporation through metal masks and photolithography have been utilized, as discussed by Tickel [5], to define the film patterns making up the device. Because complete devices over large substrate areas can be produced in a single deposition cycle with the through-mask process, this has generally become the preferred fabrication approach. This method does require, however, complex vacuum chamber fixturing and essentially perfect, high-precision, large-area, metal stencil masks. The narrow source-drain gap is defined by use of a fine-drawn wire. Fabrication of transistor arrays as described by Brody et al. [6], Brody [7], Tickel [5], and Weimer [3, 4, 8] proceeds by depositing, layer by layer, a sequence of materials through an appropriate mask. In general, 6–20 film depositions are necessary to complete the array fabrication. A common sequence for the layers consists of (1) insulator film for substrate conditioning, (2) source-drain electrode metallization, (3) semiconductor film, (4) gate insulator, (5) gate metallization, and (6) another insulator film for device passivation.

2. Applicable Semiconductor Film Materials

The key to the performance of the TFT is the polycrystalline semiconductor film. While a number of different semiconductor materials have been or could be utilized to form the device, only four have progressed to the point where they are being considered for commercial applications: CdS, CdSe, Te, and PbS.

The first successful and best known example of the TFT reported by Weimer [1] incorporated a CdS film deposited by vacuum evaporation onto a glass substrate. Other films in the CdS device are usually Al or Au electrodes and SiO_x, Al_2O_3, or CaF_2 gate insulators. According to Weimer [2], devices with transconductances as high as 25,000 μmho and routinely about several thousand μmho have been achieved. Typical operating characteristics for an enhancement mode CdS film transistor are shown in Fig. 7.2b.

As summarized by Tickel [5] and Weimer [2], the CdS films ($\sim$4000 Å in thickness) for these devices are vacuum-deposited from resistively heated sources baffled with quartz wool to prevent spitting onto heated substrates (150–200°C) at relatively high deposition rates ($\sim$5000 Å/min). These two deposition parameters have been the principal means for controlling the electrical and structural film properties. For example, the electric resistivity can be varied from 0.1 Ω cm to 10^8 Ω cm by only adjusting the substrate deposition temperature. Increasing temperature and decreasing deposition rate result in high-resistivity films. Measured Hall mobilities for CdS films producing good TFT characteristics have been quite low, i.e., only 5 cm^2/V s. The films have also been shown to consist of small crystallites ($\sim$0.1 μm) with preferential orientation (c axis normal to the substrate). Somewhat surprisingly, the fabrication of devices with a very large grain size or single-crystal CdS films produced by recrystallization with Ag has apparently not resulted in improved performance.

As reported by numerous investigators (Brody [7], Shallcross [9], Tickel [5], Weimer [4], and Zulegg [10]), CdSe has largely replaced CdS films in the fabrication of n-type semiconductor film transistors. Although selenide devices are slightly inferior in performance compared to sulfide devices, the deposition processes are less critical, resulting in better reproducibility and higher yields. The simpler deposition process has been related by Tickel [5] to the relative vapor pressures of the three involved elements. The vapor pressure of Se is much closer to that of Cd than is S, which means a much lower sensitivity to substrate temperature during film deposition. In fact, excellent selenide devices may be produced using films deposited on unheated substrates, which is highly advantageous to the through-mask fabrication approach. With a higher electron mobility, selenide transistors should also yield, according to Tickel [5], a higher frequency response for a given channel length through the reduced transit time of carriers.

The CdSe films about 1500–2000 Å in thickness have been deposited by Brody *et al.* [6], van Heek [11], Tickel [5], and Weimer [4] using techniques essentially the same as those for CdS. Again, very small-crystallite films have resulted ($\sim$0.1 μm), but the Hall mobilities of 4–25 cm^2/V s are somewhat higher than for CdS. The films are normally subjected to an air bake at about 150°C to increase the resistivity and to achieve the desired device performance.

Thin-film transistors based upon polycrystalline Te semiconductor films have also been well developed, as reported by Dutton and Muller [12], Weimer [13], and Wilson and Gutierrez [14]. Tellurium has the advantage of yielding p-type devices for complementary symmetry thin-film circuits and, because it is not a compound, the problems of dissociation during the deposition process are avoided.

Dutton and Muller [12] have formed Te films for these TFT devices by vacuum evaporation from a tungsten boat. As with CdS and CdSe, the films possess very small crystallites (~ 200–500 Å). However, these authors have shown that Au "spray" from the electrode films causes film recrystallization into very large-grain deposits (~ 2–5 μm) with improved device characteristics. The recrystallized films exhibit preferred orientation with the c axis parallel to the substrate. Transconductances as high as 40,000 μmho have been reported by Weimer [13] for p-type Te devices, which makes them excellent complementary units to n-type CdSe devices. A significant difference in Te device fabrication as compared to that of CdS or CdSe has been the use of extremely thin deposits, i.e., 150–400 Å [15]. The achievement of well-saturated enhancement mode devices with this very low-resistivity material has necessitated the use of a semiconductor layer so thin that the depletion layer produced by the gate in the neighborhood of the drain contact extends almost completely through the semiconductor. Device characteristics for a p-type Te TFT are shown in Fig. 7.3.

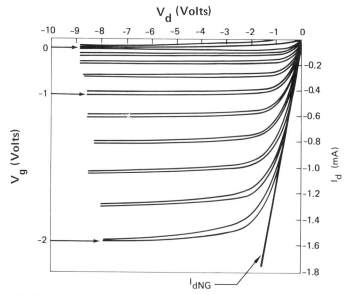

Fig. 7.3   Drain characteristics for tellurium p-type TFT. (From Weimer [13].)

Less developed than the three devices discussed previously, good-quality TFTs utilizing PbS as the semiconductor layer have been produced by Kramer [16] and Pennebaker [17]. Interest in PbS is enhanced by the fact that both *n*- and *p*-type films can be formed. In addition, a TFT device with a normally *n*-type PbS film can be inverted to *p*-type by the application of high negative gate fields. Transconductances of 30 μmho have been reported by Kramer [16] and, for good stability, the gate dielectric must be an Al_2O_3 film. As with Te, the high conductivity of PbS requires the use of very thin layers ($\sim$ 150 Å) to achieve well-saturated device characteristics. The films are deposited on unheated substrates by vacuum evaporation from tantalum or fused quartz boats in an oxygen background of 9×10^{-5} torr to control film resistivity. Kramer [16] has found the carrier concentration and electron mobility in these *n*-type films to be 2×10^{17} cm^{-3} and 10 cm^2/V s, respectively.

3. Specialized Thin-Film Transistor Device Configurations

The flexibility of a device fabrication approach based upon vacuum-deposited thin films is well demonstrated by several special-purpose TFT designs with important device application potentials. For example, a non-volatile CdSe memory TFT has been developed by Yu *et al.* [18] and Yu [19], which could be very useful in the construction of large-area solid-state displays. The structure of this floating-gate memory device is depicted in Fig. 7.4 and it differs from the standard CdSe TFT in that the CdSe film is extremely thin (100–200 Å) and the gate dielectric is an Al_2O_3–Al–SiO_x sandwich. The Al film is only 5–30 Å in thickness and consists of metallic clusters. Charging and discharging of these clusters with the gate field results in the memory action, and on/off current ratios greater than 1000 are achieved.

In another modification reported by O'Hanlon and Haering [20] to the conventional CdS TFT device, the gate insulator thickness was increased from 1000 to 10,000 Å and the source-drain gap width increased from 7.5

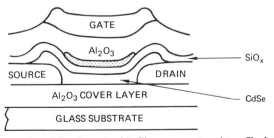

Fig. 7.4 Cross section of floating-gate thin-film memory transistor. Shaded area: floating gate. (From Yu *et al.* [18].)

to 250 μm. With these changes, the device control voltages were increased to the order of 50 V, which is ideally suited to electroluminescent cell applications.

7.3 DIODES

As demonstrated by the results of Weimer [2, 4], field-effect and Schottky-barrier diodes are the two prominent device configurations used in thin-film integrated circuits. Field-effect devices are especially attractive, since they can be formed at the same time and by compatible procedures as thin-film triodes. In fact, a well-known version described by Weimer [2, 4] of the field-effect diode simply ties the gate electrode to the drain. The more common configuration is, however, constructed by omitting the gate

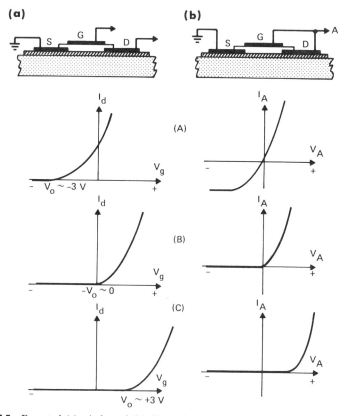

Fig. 7.5 Expected (a) triode and (b) diode characteristics of TFTs with interconnected gate-drain electrodes for different valves of V_0. (From Weimer [2].) (A) $V \sim -3V$, (B) $V_0 \sim 0V$, (C) $V_0 \sim +3V$.

electrode and separating the inner edge of the drain electrode from the semiconductor with a thin insulating film. Satisfactory diode performance can be obtained by controlling the value of V_0 (threshold gate voltage) during the device fabrication process. The effect predicted by Weimer [2] of V_0 on diode and triode behavior is shown in Fig. 7.5. With V_0 values near zero, CdS and CdSe field-effect diodes have been produced with rectification ratios of 10^5.

Thin-film Schottky-barrier diodes are formed by contacting the semiconductor film with a metal or other material, which produces a potential barrier at the contact. The second diode electrode is a metal which makes a good ohmic contact to the semiconductor layer. Weimer [4] has reported forming a diode utilizing CdS as the semiconductor and Te as the blocking contact. The diode construction was the sequential deposition of films of Au, In, CdS, Te, and Au onto an unheated substrate. The Au films acted as low-resistance electrodes (Au forms an ohmic contact to Te), while In was used to ensure an ohmic contact to the CdS film. These diodes showed rectification ratios greater than 10^4, with forward current densities of tens of amperes per square centimeter.

7.4   PHOTOCONDUCTORS

Photoconductivity (an increase in electric conductivity with the absorption of light or other suitable radiation) occurs in all semiconductor materials. Tables of photoconductors have been assembled by Gorlich [21], and photoconductivity has been the subject of many reviews, such as those by Bube [22] and by Rose [23]. While the total number of photoconductor materials is large, Bube [24] states that less than 10 are in widespread commercial usage. These include Si, Ge, CdS, CdSe, GaAs, InSb, and the lead salts. Photoconductors may be in the form of single crystals, thick-sintered pellets, or thin-sintered, evaporated, chemically deposited, or sputtered layers. Sulfide, selenide, and lead salts are the predominant polycrystalline thin-film semiconductor materials. Indeed, defects inherent to the polycrystalline structures play an important role in achieving successful device performance characteristics.

Photoconductors are used in radiation detectors, control elements, computer applications, television cameras, electrophotography, x-ray intensifiers, and picture reproduction and display. The selection of a semiconductor of a particular material for a specific application depends largely upon the spectral response of the semiconductor, as summarized in Fig. 7.6.

Photoconductivity in semiconductors may be due to either intrinsic or extrinsic excitation. Intrinsic photoconductivity results when the incident radiation has energies greater than the band-gap energy, thus creating

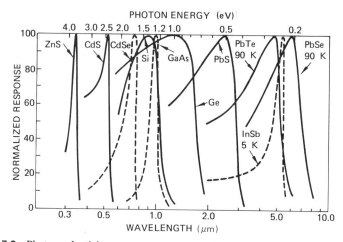

Fig. 7.6 Photoconductivity response curves as a function of optical wavelength for a number of photoconductor materials. (From Bube [22].)

electron–hole pairs and a higher conductivity. Extrinsic behavior results when free carriers are photoexcited from impurity centers with energy levels within the forbidden energy gap. Intrinsic mode devices generally operate in the wavelength region from 0.1 to 1.0 μm, while extrinsic photoconductors have achieved most prominence as infrared detectors. In either case, however, high photosensitivity is achieved by intentionally incorporating impurities into the base semiconductor material. These impurities produce photoconductive devices with high gain, i.e., a large number of charge carriers which pass between the device electrodes for each photon absorbed. Bube [22, 24] has shown that photoconductivity gain may be expressed as

$$G = \tau / t_{\text{tr}} = \mu V / L^2 \qquad (7.3)$$

where τ is the lifetime of a free carrier, t_{tr} the transit time for this carrier, μ the carrier mobility, V the applied voltage, and L the electrode spacing. Sensitizing impurities trap the minority carriers, causing a reduction in the recombination processes and long majority carrier lifetimes. According to Bube [22, 24], Vecht [25], and Weimer [4], the lifetimes may range from 10^{-4} to 10^{-2} s for a sensitive photoconductor with gains exceeding 10^4 electrons per photon.

Although polycrystalline thin-film photoconductors have not achieved gains as high as single-crystal samples, the fact that large, uniform-area, inexpensive devices can be produced is of much practical interest. This is especially true if a widely used photoconductor film deposition process such as vacuum evaporation is considered. Weimer [4] has found, for example, that the ability to form complex circuits in a single pump-down of a vacuum chamber offers significant advantages in the elimination of defects, one of the major problems in the fabrication of photoconductor components such as image sensors. Evaporated film structures vary from amorphous to polycrystalline, depending upon the makeup of the vacuum environment, rate of deposition, and substrate temperature (Bube [22]). Chopra [26] has reported that films of the important II–VI and IV–VI photoconductor compounds deposited at room temperature generally have very fine-grained structures, have low electric resistivities, and exhibit little or no photoconductivity. Increasing the substrate temperature or subjecting the films to a postdeposition anneal results in an increase in grain size, resistivity, and photoconductive response. In the high-gain devices of more general interest, the anneal or deposition process must also provide for the incorporation of certain sensitizing impurities.

1. Films of II–VI Compounds

Historically, as described by Bube [22, 24], Seib and Aukerman [27], and Vecht [25], the preferred method for producing photoconductor devices for the visible spectral region with the II–VI compounds CdS, CdSe, and CdSSe is a sintering process. Evaporated films are, however, more suitable for advanced technology applications such as high-quality image sensors, image intensifiers, and high-resolution cathode ray screens, according to Chopra [26] and Vecht [25]. The vacuum deposition procedures for these 1–5 μm-thick films are quite similar to those used in the fabrication of TFTs, as discussed previously. It is mainly in the postdeposition processing for photoconductivity sensitization or activation that the two device fabrication approaches differ. An excellent review of the deposition and activation processes for II–VI photoconductor films has been provided by Vecht [25].

Photoconductivity activation in CdS and CdSe films is typically accomplished by incorporating a group I element—usually Cu. As indicated by Bube [22], Seib [27], and Vecht [25], Cu provides charge compensation and a hole-trapping level in these *n*-type semiconductor materials. Another important function of Cu demonstrated by Dresner and Shallcross [28] is to improve greatly the film crystallinity both in terms of grain size and crystallite orientation. A coactivator from group VII (usually Cl) is also frequently incorporated. The role of coactivators is still uncertain, but it has been reported by Bube [22] and Vecht [25] that they are donors and influence diffusion of the activator into the host lattice. Impurity concentrations are generally in the 10–1000 ppm range. One of the best doping methods, developed by Gans [29], consists of placing the film in a powder matrix of the same semiconductor material doped with a high concentration of the desired impurities and then heating the package to the required diffusion temperature (450°–650°C). Films prepared in the above manner, as reported by Dresner and Shallcross [28] and Vecht [25], exhibit high light-to-dark resistivity ratios, high photoconductivity gain, and good time constants or speed of response. Vecht [25] has also noted that imperfections associated with the polycrystalline film structure can lead to a faster response and make film photoconductive cells more useful than single crystals.

Another desirable feature of the CdS and CdSe material systems for photoconductor device applications, discussed by Gorlich [21], is the ease with which the two semiconductors can be mixed to form solid solutions with spectral responses intermediate to those of the two end members. The value of this flexibility has been demonstrated by Weimer et al. [30] in the construction of visible-spectrum photoconductor arrays (which also included TFTs and thin-film diodes) for experimental television cameras.

2. Films of IV–VI Compounds

As reported by Bode [31], Bube [22, 24], and Gorlich [21], the lead compounds PbS, PbSe, and PbTe (especially PbS) have long been used for thin-film infrared detectors. Moreover, films of PbO have been found by Wang et al. [32] to be competitive with amorphous Se and Sb_2S_3 as the photoconductor layer in the vidicon camera developed by Weimer et al. [33].

The sulfide, selenide, and telluride lead salts are polycrystalline films with crystallite sizes ranging from 100 Å to several microns, depending upon the method of fabrication. They are usually intrinsic detectors with a spectral response, as indicated in Fig. 7.6. Even though selenide and telluride are more responsive at longer wavelengths, they have not been as fully developed nor, as reported by Bode [31], do they have as high a figure of merit for detectivity D^* as PbS. The parameter D^* has evolved to indicate directly how effective the detector is in distinguishing between a small photoconductive signal and random noise due to the detector and its environment. Bube [24] has expressed detectivity by the equation

$$D^* = (A\,\Delta f)^{1/2}/(\text{NEP}) \qquad (7.4)$$

where A is the working area of the detector, Δf the frequency bandwidth for the measuring conditions, and NEP the noice equivalent power (the flux in infrared radiation which gives a signal-to-noise ratio of unity). Bode [31] and Bube [24] report values for D^* in the 10^9–10^{11} cm $Hz^{1/2}$/W region can be obtained by cooling the detectors to minimize the high, thermally excited dark conductivity and by chopping the incoming infrared radiation.

As summarized by Bode [31] and Bube [22], the lead salt detectors are prepared primarily by chemical deposition, but vacuum evaporation is also used with good results. For chemical deposition of PbS onto a variety of substrates, the fundamental chemical reaction is that between thiourea and Pb acetate in a basic solution. Crystallite sizes in these micron-thick films are generally in the 0.1–1.0 μm range.

Regardless of the film deposition method, the most common photoconductivity sensitizing agent for the lead salt detectors is oxygen, as documented by Bode [31], Bube [22], and Petritz [34]. Oxygen may be introduced during the film deposition by vacuum evaporation with an oxygen leak or by a postdeposition bake in air or under reduced pressure of oxygen. Bake temperatures are typically 300°–400°C. The role of oxygen and the photoconductivity mechanism itself in these layers is still the subject of much debate. Sensitizing centers, intercrystalline oxide barriers, p–n junctions, and other structures have all been suggested. It may also be of interest to note that polycrystalline thin-film structural effects initially observed and investigated by Petritz [34] on PbS film photoconductors have been applied by a number of authors to other film materials and devices including the TFT [2, 5, 11, 28, 35–39].

The results of Wang *et al.* [32] indicate that the remaining lead compound to be discussed, namely, PbO, differs from the above infrared detectors in its large band-gap energy (1.94 eV), photoconductivity threshold wavelength of 0.65 μm, and high electric resistivity (about 10^{12} Ω cm). These properties are appropriate for vidicon camera applications, and polycrystalline PbO thin films have been developed for this purpose. In comparison to the more common Sb_2S_3 and amorphous Se film layers, PbO film devices have been reported to be of lower dark current, higher resolution, higher sensitivity, and lower lag. The film deposition method consisted of heating tetragonal PbO (red form) in a residual atmosphere (10^{-2} torr) of O_2 and H_2O. The O_2 served to prevent formation of free Pb and to maintain film stoichiometry–conductivity type. Water was believed to be the transporting agent and also to control the film crystallite size which, in turn, determined the resolution of the camera tube target. Crystallites were in the form of upright platelets with dimensions of approximately $1 \times 0.5 \times 0.05$ μm to $2 \times 1 \times 0.1$ μm. Like the Sb_2S_3 films, these PbO deposits contained a large amount of porosity, i.e., 45%–60%.

For vidicon operation, the PbO film is deposited onto an electrically conducting tin oxide-coated glass substrate. It is then exposed to a scanning electron beam on one side (film surface) and to the incident light on the other. The surface of the high-resistivity photoconductor is charged by the scanning electron beam, while a potential of 10–30 W is applied to the conducting coating contacting the opposite side of the film layer. Exposure to light permits charge to flow, causing the illuminated areas to become slightly positively charged. The electron beam then neutralizes this charge and thus creates the signal containing the information about the scene brightness.

7.5   LUMINESCENT THIN FILMS

According to Goldberg [40], serious consideration of luminescent thin films for device applications began in the 1950s. The driving force for this work was the promise of unusual or improved large-area devices such as high-resolution, high-contrast cathode ray tubes, optical elements of high resolving power, and electroluminescent (EL) displays. As indicated above, the two luminescence phenomena of commercial significance for thin-film materials are cathodoluminescence (excitation by impact of electrons of intermediate energies, i.e., 1–30 keV) and electroluminescence (excitation by application of electric potential). Thin films offer possible advantages over conventional powder phosphors in the area of better contrast and resolution but have not experienced extensive usage. The major reasons which have limited exploitation of thin films are high cost, low efficiency and, especially in the case of EL devices, short operation lifetimes. Recently, however, substantial progress has been made in the development of stable devices with improved brightness, as shown by the results of Fugate [41], Inoguchi et al. [42], and Inoguchi and Mito [43]. The compatibility of these devices with advanced technology TFTs for large-area, flat-screen displays has prompted a renewal of interest in the electronics industry.

Polycrystalline thin-film phosphors of practical importance are the II–VI semiconductor compounds ZnS, ZnSe, and CdS. Of these ZnS is by far the predominant material, although it is frequently mixed with CdS or ZnSe to achieve certain desired characteristics. The films are processed to produce high-resistivity material containing special activator impurities for good luminescent characteristics. Thus, luminescent thin films are seen to have features in common with photoconductor films. While there are many similarities between luminescence and photoconductivity, there are also major differences, as discussed by Bube [22]. Good luminescent materials may not necessarily make good photoconductors, and vice versa.

In luminescence, the recombination of activated carriers is a radiative process. As Beam [44] and Bube [22] have related, by using wide-band-gap semiconductors (ZnS, $E_g = 3.7$ eV; ZnSe, $E_g = 2.6$ eV; CdS, $E_g = 2.4$ eV), radiative transitions of less than the full band-gap energy fall within the optical range from about 1.8 to 3.0 eV. Activator impurities in these materials have energy states near the valence band and serve as hole traps or recombination centers. Many phosphors also contain coactivator im-

TABLE I

Zinc Sulfide Thin-Film
Electroluminescent Activators

Activator	Emission wavelength
Cu, Cl	Blue
TmF_3	4750 Å
PrF_3	5000 Å
ErF_3	5210 Å
TbF_3	5430 Å
Cu, Cl	Green
DyF_3	5630 Å
Mn	5800 Å
SmF_3	6500 Å
NdF_3	6000–8900 Å

purities which have energy levels near the conduction band and act as electron traps.

The major criterion for individual activator selection is the desired emission spectrum. There is, however, some influence of activator type and concentration on device efficiency according to Goldberg [40] and Vecht [25]. A list of activators in ZnS compiled by Soxman and Ketchpel [45] is presented in Table I. The rare earth materials are a more recent development and are not as extensively used as Cu–Cl and Mn. Except for Cu–Cl, emission produced by all the activators in Table I has been attributed by Bube [22], Koller and Coghill [46], and Soxman and Ketchpel [45] to transition between states of the metal activator and not to transition from the conduction band to the activator state. A convenient method of shifting the spectral emission curve, for example, to produce a better match with the sensitivity of the human eye, is to mix the ZnS with CdS or ZnSe, as reported by Goldberg [40].

1. Cathodoluminescent Films

Polycrystalline, cathodoluminescent thin films of ZnS 1–5 μm in thickness may be prepared by vacuum evaporation, spraying, vapor phase reaction, or chemical vapor transport. Vacuum evaporation is the most flexible approach and has the greatest potential for future advanced technology devices. Methods for incorporating the activating impurities (Cu–Cl, Mn, Ag) have varied considerably but, as discussed by Goldberg [40] and Vecht [25], generally fall into one of two categories, i.e., a one-step or a two-step process.

In the two-step approach, the ZnS film is deposited at room temperature and then activated by baking to diffuse in the impurities. Sources of impurities may be a very thin deposited film, doped powders, or a gas atmosphere. Baking the ZnS films (450°–650°C) not only leads to the incorporation of impurities but, as found by Kollor and Coghill [46], crystallization of the initially amorphous film. The development of a good crystalline structure is essential for strong luminescence. Grain size in the crystallized films is in the 0.1–1.0 μm region.

The one-step fabricated method described by Koller and Coghill [46], Nickerson and Goldberg [47], Reams [48], and Soxman and Ketchpel [45] utilizes deposition onto heated substrates (250°–650°C). Impurities in these now crystalline films have been incorporated by evaporating doped ZnS powders, coevaporation, and/or depositing under an HCl residual gas pressure. Manganese is an especially attractive dopant, since it can serve as the sole activator, is used in high concentrations (1%–5%) which lessens material purity requirements, and is less sensitive to film crystalline perfection since the luminescence is between levels of the Mn center and not from the conduction band to the center. Vaporization of either the metallic element or the compound MnS has proved successful in preparing activated ZnS:Mn films.

Cathodoluminescence colors and concentrations for the various activators are as follows:

(1) Cu–Cl with excess Cu is blue, but the excess Cl is green; typical concentrations are 0.01 to about 1%;

(2) Mn (1%–5%) produces a yellow-orange emission; and

(3) Ag–Cl (0.01%) results in blue emission. Luminous efficiencies for thin-film phosphors are less than 10%, which is less than half that of a good powder material according to Vecht [25].

To counteract their lower efficiencies and higher costs, thin films offer potentially better resolution, contrast, and advanced device configurations. For example, Goldberg [40] and Vecht [25] indicate that transparent thin films on a blacklight-absorbing layer should permit good contrast even with high background illumination levels, and multilayer film structures permit the emission color to be varied by adjusting the beam voltage.

2. Electroluminescent Films

Both ac- and dc-operated EL devices based upon activated ZnS phosphors have been produced in powder and thin-film form. The ac-driven thin-film structures have been found by Goldberg [40], Howard [49], Kazan [50], and Soxman and Ketchpel [45] to possess several important advantages, which should lead to the predominance of this approach

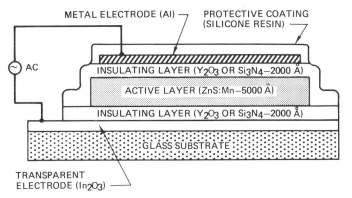

METAL ELECTRODE (Al) ⌐   PROTECTIVE COATING
(SILICONE RESIN) ⌐

AC

INSULATING LAYER (Y_2O_3 OR Si_3N_4—2000 Å)

ACTIVE LAYER (ZnS:Mn—5000 Å)

INSULATING LAYER (Y_2O_3 OR Si_3N_4—2000 Å)

GLASS SUBSTRATE

TRANSPARENT
ELECTRODE (In_2O_3)

Fig. 7.7   Schematic structure of a thin-film EL device. (From Suzuki *et al.* [51].)

in future applications. Voltages and frequencies for the devices are 100–300 V and 0.3–5 kHz, respectively. Although a split-electrode configuration has recently been reported by Fugate [41], the usual device structure consists of a sandwich of electrode, phosphor, and dielectric film deposited onto a glass substrate. The purpose of the dielectric layer is to afford protection against voltage breakdown during operation. Initially only one dielectric layer was used, but advanced devices as reported by Inoguchi *et al.* [42], Inoguchi and Mito [43], Soxman and Ketchpel [45], and Suzuki *et al.* [51] now include two layers and, in some instances, a blacklight-absorbing film. In the newer structures, the deposited layered sequence is a transparent conducting electrode (SnO_2 or In_2O_3), a dielectric film (Y_2O_3, GeO_2, Si_3N_4, HfO_2, Al_2O_3, SiO_2), activated phosphor, a second dielectric film or black layer (As_2S_3), and a metal electrode film (Al). A device of this type is depicted in Fig. 7.7.

Even more than with cathodoluminescent ZnS films, the element Mn in concentrations of 1%–5% has been applied for phosphor activation in EL thin-film devices. Techniques for incorporating the Mn and the ZnS film deposition are basically the same as for cathodoluminescent films. Several variations on the conventional approach to be mentioned are the use of sputtering for ZnS film deposition and activation [52, 53], rare earth activator materials [45, 52, 54], and Mn activation by ion implantation [55].

As summarized by Kazan [50], substantial progress has been shown by recent ZnS:Mn EL thin-film devices in the areas of stability and brightness, but they are still only about $\frac{1}{10}$ as bright as cathodoluminescent powder phosphors. Their characteristics should be well suited for displays, but further improvements in brightness, available colors, and cost must be achieved before this technology can make the long-anticipated major impact on the television industry.

7.6   FUTURE EXPECTATIONS

The continued emergence and growth of the described thin-film devices in the electronics marketplace will most likely be determined by economic rather than performance factors. As a consequence of a quite extensive research and development effort involving these devices over a number of years, their performance relevant to many critical parameters has achieved levels competitive with the devices they have been designed to either replace or extend in capability. It appears that the principal concern regarding future implementation is that of manufacturing cost. At the present time, higher fabrication costs have effectively negated the technical advantages demonstrated or predicted by the thin-film approach. Electronic systems can be expected, however, to continue expanding in size, complexity, and performance requirements. Further device improvements and the application of advanced film deposition–processing–analysis equipment to the establishment of lower-cost, higher-yield production methods could tip this economic balance in favor of thin films. The combination of thin-film devices, e.g., TFTs and diodes with photoconductors for large-image sensor arrays and TFTs with EL devices for displays and flat-screen television, could result in the necessary cost–performance advantage. The semiconductor industry with its constant evolution and demise of a device technology makes any forecast somewhat suspect, but it is reasonable to assume that devices incorporating polycrystalline semiconductor thin films will provide a significant contribution in one or several application areas.

REFERENCES

1. P. K. Weimer, *IRE Trans. Electron. Devices* **ED8**, 421 (1961).
2. P. K. Weimer, "Physics of Thin Films" (R. Thun and G. Hass, eds.), Vol. 2, pp. 147–192. Academic Press, New York, 1964.
3. P. K. Weimer, *Proc. IEEE* **52**, 1479 (1964).
4. P. K. Weimer, "Handbook of Thin Film Technology" (L. I. Maissel and R. Glang, eds.), pp. 20-1–20-18. McGraw-Hill, New York, 1970.
5. A. C. Tickel, "Thin Film Transistors." Wiley (Interscience), New York, 1969.
6. T. P. Brody, D. H. Davies, F. C. Luo, E. W. Greeneich, G. D. Dixon, and J. A. Asars, Flat Plate Displays. Final Rep. AFAL-TR-74-301 (1974).
7. T. P. Brody, *Proc. SID* **17/1**, 39 (1976).
8. P. K. Weimer, *Proc. IEEE* **54**, 354 (1966).
9. F. V. Shallcross, *Proc. IEEE* **51**, 851 (1963).
10. R. Zuleeg, *Solid State Electron.* **6**, 645 (1963).
11. H. F. van Heek, *Solid State Electron.* **11**, 459 (1968).
12. R. W. Dutton and R. S. Muller, *Proc. IEEE* **59**, 1511 (1971).
13. P. K. Weimer, *Proc. IEEE* **52**, 608 (1964).
14. H. L. Wilson and W. A. Gutierrez, *Proc. IEEE* **55**, 415 (1967).

15. J. C. Anderson, "Active and Passive Thin Film Devices" (T. J. Coutts, ed.), pp. 207–245. Academic Press, New York, 1978.
16. G. Kramer, Thin-Film Transistor Stability. Final Rep. 74-0265/AESC 5014 (1975).
17. W. B. Pennebaker, *Solid State Electron.* **8**, 509 (1965).
18. K. K. Yu, T. P. Brody, and P. C. Y. Chen, *Proc. IEEE* **63**, 826 (1975).
19. K. K. Yu, *IEEE Trans. Electron Devices* **ED-24**, 591 (1977).
20. J. F. O'Hanlon and R. R. Haering, *Solid State Electron.* **8**, 509 (1965).
21. P. Gorlich, *Adv. Electron. Electron. Phys.* **14**, 37 (1961).
22. R. H. Bube, "Photoconductivity of Solids." Wiley, New York, 1960.
23. A. Rose, "Concepts in Photoconductivity and Applied Problems." Wiley, New York, 1963.
24. R. H. Bube, *Trans. Metall. Soc. AIME* **239**, 291 (1967).
25. A. Vecht, "Physics of Thin Films" (R. Thun and G. Hass, eds.), Vol. 3, pp. 165–210. Academic Press, New York, 1966.
26. K. L. Chopra, "Thin Film Phenomena." McGraw-Hill, New York, 1969.
27. D. H. Seib and L. W. Aukerman, *Adv. Electron. Electron. Phys.* **34**, 95 (1963).
28. J. Dresner and F. V. Shallcross, *J. Appl. Phys.* **34**, 2390 (1963).
29. F. Gans, *Bull. Sci. AIM* **11**, 897 (1953).
30. P. K. Weimer, G. Sadasiv, J. E. Meyer, L. Meray-Horvath, and W. S. Pike, *Proc. IEEE* **55**, 1591 (1967).
31. D. E. Bode, "Physics of Thin Films" (R. Thun and G. Hass, eds.), Vol. 3, pp. 275–301. Academic Press, New York, 1966.
32. C. C. Wang, K. H. Zaininger, and M. T. Duffy, *RCA Rev.* **31**, 728 (1970).
33. P. K. Weimer, S. V. Forgue, and R. R. Goodrich, *Electronics* **23**, 70 (1950).
34. R. L. Petritz, *Phys. Rev.* **104**, 1508 (1956).
35. A. Amith, *J. Vac. Sci. Technol.* **15**, 353 (1978).
36. Z. T. Kuznicki, *Solid State Electron.* **19**, 894 (1976).
37. Z. T. Kuznicki, *Thin Solid Films* **33**, 349 (1976).
38. C. A. Neugebauer, *J. Appl. Phys.* **39**, 3177 (1968).
39. A. Waxman, V. E. Henrich, F. V. Shallcross, H. Borkan, and P. K. Weimer, *J. Appl. Phys.* **36**, 168 (1965).
40. P. Goldberg, "Luminescence of Inorganic Solids." (P. Goldberg, ed.). Academic Press, New York, 1966.
41. K. O. Fugate, *IEEE Trans. Electron Devices* **ED-24**, 909 (1977).
42. T. Inoguchi, C. Suzuki, and S. Mito, *Jpn. Electron. Eng.* **118**, 30 (1976).
43. T. Inoguchi and S. Mito, "Electroluminescence" (J. Pankove, ed.), pp. 197–210. Springer-Verlag, Berlin and New York, 1977.
44. W. R. Beam, "Electronics of Solids." McGraw-Hill, New York, 1965.
45. E. J. Soxman and R. D. Ketchpel, Final Rep., JANAIR 720903 (1972).
46. L. R. Koller and H. D. Coghill, *J. Electrochem. Soc.* **107**, 973 (1960).
47. J. W. Nickerson and P. Goldberg, *Nat. Symp. Vac. Technol. Trans.* **10**, 475 (1963).
48. J. P. Reams, *Nat. Symp. Vac. Technol. Trans.* **6**, 215 (1959).
49. W. E. Howard, *IEEE Soc. Informat. Display, Biennial Display Conf.* **12**, 9 (1976).
50. B. Kazan, *Proc. S.I.D.* **17/1**, 23 (1976).
51. C. Suzuki, T. Inoguchi, and S. Mito, *Informat. Display* **13**, 14 (1977).
52. R. A. Buchanan, T. G. Mapel, H. N. Bailey, and R. V. Alves, Final Rep. LMSC-D315254, No. AD-758-760 (1973).
53. J. J. Hanak and P. N. Yokum, Final Rep. PRRL-73-CR-10, No. AD-766-741 (1973).
54. Y. S. Chen, M. V. DePaolis, Jr., and D. Kahng, *Proc. IEEE* **58**, 184 (1970).
55. T. Takagi, I. Yamada, A. Sasaki, and T. Ishibashi, *IEEE Conf. Display Devices* **8**, 51 (1972).

8 | Thin-Film Photovoltaic Devices

ALLEN ROTHWARF,* JOHN D. MEAKIN,
and ALLEN M. BARNETT[†]

Institute of Energy Conversion
University of Delaware
Newark, Delaware

*Present address: Electrical Engineering Department, Drexel University, Philadelphia, Pennsylvania.

[†] Present address: Electrical Engineering Department, University of Delaware, Newark, Delaware.

8.1 INTRODUCTION

Reliable and low-cost renewable sources of energy, particularly electric energy, are being sought utilizing many different conversion systems. Thin-film solar cells represent one such technology which has recently attracted increased attention and which is now being actively pursued using many different material systems. While single-crystal silicon solar cells have proved to be reliable in both space and certain terrestrial applications, there is little likelihood that the cost per watt of these cells can be reduced to the point that widespread use will be possible. Thin-film polycrystalline solar cells will require less material and are more suited to mass-production techniques and are therefore viewed as a viable low-cost option. An extensive literature [1–14] covering both single-crystal and polycrystalline thin-film solar cells exists already. In this chapter we have attempted to give an overview of the present situation in research and development to provide a basis for forecasting the future of low-cost cells.

1. Generic Solar Cell

The solar cell can be conveniently divided into an absorber–generator region in which the light is absorbed and electron–hole pairs generated, and a collector–converter region in which the generated carriers are separated and passed through to the contact. One can define the photovoltaic effect as the generation of carriers by the action of light and the separation of these carriers from the region in which they were generated by the presence of a barrier to reverse flow and recombination.

2. Definition of a Thin Film

There are many techniques which build up films essentially layer by layer, and for the purposes of this chapter any film grown by such a technique will be called a thin film. These techniques are described below and also in other chapters. For our purposes thin films generally range from an order of several thousand angstroms to as much as 30 μm in thickness.

3. Advantages of Thin Films

The main interest in studying thin-film photovoltaic devices is that they represent a potential low-cost mass-production technique for producing high-efficiency solar cells. The total requirements for material are reduced in proportion to the thickness. Because the active semiconductor layers are thin, more imperfections and impurities can be tolerated than in the thicker single-crystal cells. Thin-film polycrystalline solar cells are leading contenders for a large-scale technology for the economic conversion of sunlight to electricity.

4. Problems of Thin-Film Cells

Thin-film cells generally are polycrystalline or amorphous materials. The polycrystalline nature of the materials introduces internal surfaces in the form of grain boundaries. These grain boundaries can degrade current generation, voltage, and the stability of the cells. The elimination or amelioration of these effects is the main challenge in perfecting thin-film photovoltaic materials. The method of deposition leads, in many cases, to nonstoichiometry and to efficiency-limiting surfaces. Postdeposition recrystallization of the as-deposited texture may also create problems related to material orientation [15]. A class of problems arises because thin films must be deposited on a suitable substrate. For low-cost applications the substrate must itself be of low cost. In addition it must make ohmic contact with the semiconductor components, have suitable thermal and optical properties, and match the adjacent layers in lattice constant and thermal expansion. The successful development of a thin-film solar cell requires the understanding and control of multiple interactions between complex materials.

5. Types of Cells

There are three major types of cells under active study. The first cells to reach practical application were homojunctions. They contain the same material doped to produce the *p*- and *n*-sides of the cell. Heterojunctions constitute a second major classification of cells. They consist of dissimilar materials and as a consequence have an interface with a potential for problems of lattice mismatch and differences in electron affinity. The third

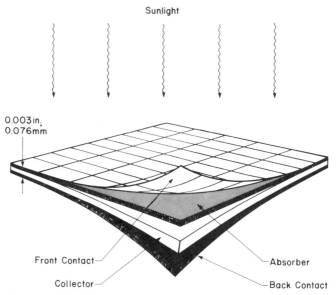

Fig. 8.1 Generic solar cell illustrating the necessary components of a thin-film solar cell.

class of devices consists of Schottky diodes which are either thin metal films or transparent conductors on an absorbing semiconductor. The equations which describe the operation of these cells are given below and will be discussed in detail in subsequent sections.

Figure 8.1 is a cross section of a generalized solar cell. The usual two semiconductor layers are shown, namely, the absorber–generator and the collector–converter. Cells are conventionally known as front wall or back wall, respectively, when the light passes through the absorbing material before reaching the junction, or vice versa. Each design has specific advantages depending on the component materials and method of cell fabrication. In Fig. 8.2, we have illustrated the energy band diagrams for homojunction, heterojunction, and Schottky-barrier cells. They are all similar in that a diffusion voltage and a barrier to carrier flow exist at the junction between either the two types of material in a homojunction, between the different materials in a heterojunction, or between the semiconductor and metal or transparent conductor in Schottky cells.

The current–voltage relation of an illuminated solar cell is convention-ally described as the superposition of the dark diode relation $j_D(V)$ and the light-generated current density J_L. In general exact superposition is not found because of light-induced changes in the charge state of various defect levels and the presence of series resistance [16]. In addition, J_L, which has traditionally been identified with the short-circuit current den-

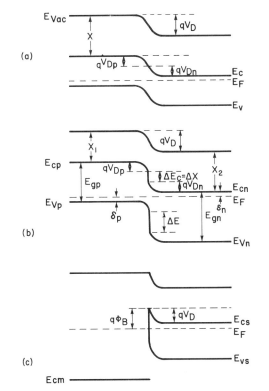

Fig. 8.2   Band diagrams of three types of solar cells: (a) homojunction, (b) heterojunction, and (c) Schottky diode.

sity, will generally be voltage dependent [16, 17]. For the current–voltage relations, we will write the equations which hold for the single-crystal versions of the cells and point out the quantities that are affected by grain boundaries and other effects.

The overall equation for the p–n junction diode solar cell, neglecting linear series and shunt resistance terms, is

$$j = \sum_i j_{0i}\left[\exp(\beta_i V) - 1\right] - j_{\mathrm{L}} \tag{8.1}$$

where j_{0i} are the coefficients for various current-flow mechanisms and β_i the corresponding factors in the diode exponential; $\beta_i = q/A_i kT$, j_{L} is the light-generated current density, and V the voltage.

For a homojunction, two terms are present in the sum. The first term involves diffusion and bulk recombination, giving $A_1 = 1$ and

$$j_{01} = q\left(\frac{D_n}{L_n}n_{0p} + \frac{D_p}{L_p}p_{0n}\right) \tag{8.2}$$

where q is the magnitude of the electronic charge, D_n the electron diffusion coefficient, and L_n the diffusion length of electrons in the p-type material; D_p and L_p are the corresponding quantities for holes in the n-type material; n_{0p} and p_{0n} are the equilibrium density of electrons and holes in p-type and n-type material, respectively.

The diffusion coefficients and lengths are related to basic material properties through the equation $D = kT\mu/q$, and $L^2 = kT\mu\tau/q = D\tau$, where μ is the mobility and τ the recombination lifetime of the minority carrier. For the second term $A_2 = 2$ and

$$j_{02} = qn_i w/\tau \qquad (8.3)$$

with $n_i = (N_c N_v)^{1/2} \exp - (E_g/2kT)$ and w the width of the space charge region.

For heterojunctions in addition to the current mechanisms present in homojunctions, there is also an interface recombination path which gives $A_3 = 1$ and [18]

$$j_{03} = qN_c S_I \exp - \phi/kT \qquad (8.4)$$

where S_I is the interface recombination velocity. Here S_I is approximately $v_{th}\sigma \Delta a/a^3$, where v_{th} is the thermal velocity of electrons (or holes), σ the capture cross section, Δa the difference in lattice constant between the two materials, a the average lattice constant in the plane of the junction, and ϕ an activation energy given by the relation (see Fig. 8.2)

$$\phi = E_{g1} - \delta_1 - qV_{D1} - (X_2 - X_1) \qquad (8.5)$$

For a Schottky diode an expression for j_0 is

$$j = A^* T^2 \exp - (\phi_B/kT) \qquad (8.6)$$

with $A^* = 120 m^*/m_0 \text{ A/cm}^2 K$.

In Fig. 8.3 the current–voltage relation in the light and dark is illustrated. The product of current and voltage is power, and the maximum power a solar cell can deliver occurs at a voltage V_{mp} and current I_{mp} schematically illustrated by the area under the maximum power rectangle in Fig. 8.3. The quantities of importance are the short-circuit current $I_{sc} \simeq j_L A$ (A is the cell area), the open-circuit voltage V_{oc}, and the fill factor FF. The fill factor FF is a measure of the squareness of the current–voltage curve and is defined as the product of the current and voltage at the maximum power point divided by the product of the open-circuit voltage and short-circuit current. In terms of these quantities, the energy conversion efficiency of a solar cell η is given by

$$\eta = I_{sc} V_{oc} FF/P_{in} \qquad (8.7)$$

where P_{in} is the power contained in the incident light.

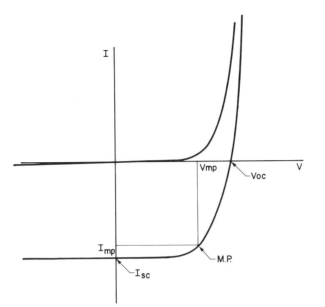

Fig. 8.3 The current–voltage curve for a solar cell illustrating the key parameters: short-circuit current I_{sc}, open-circuit voltage V_{oc}, and the maximum power point (M.P.).

If only one mechanism dominates at V_{oc}, setting $j = 0$ in Eq. (8.1) gives us immediately an expression for V_{oc}:

$$V_{oc} = (1/\beta)\ln\left[(j_L/j_0) + 1\right] \tag{8.8}$$

For homojunctions such as silicon, j_0 is given by Eq. (8.2), and for heterojunctions such as CdS–Cu$_2$S, Eq. (8.4) holds. For Schottky cells Eq. (8.6) may hold if no interfacial layer forms between the metal and semiconductor.

6. Materials

A host of materials have been and are being studied for thin-film solar cells. The most advanced thin-film solar cell and the one which we will use as an example in the following discussion is based on the CdS–Cu$_2$S heterojunction. Other materials under active development include amorphous silicon in the form of Schottky devices, CdTe as a homojunction and in various heterojunction combinations, polycrystalline silicon and GaAs, CuInSe$_2$, and InP with CdS. Each material and cell design has its own unique problems, but there is a set of problems common to all thin-film cells which we will discuss using the CdS–Cu$_2$S system to illustrate some of the more important ones.

7. Formation Techniques

Evaporation (physical deposition, molecular beam epitaxy), sputtering, chemical vapor transport, and chemical spray deposition represent a few of the major techniques. A more detailed discussion of these techniques and others can be found in Chapter 1 and also in the work of Amick et al [19]. Most of the techniques treated in Chapter 1 are of relevance to this chapter because they are potentially applicable to large-scale, low-cost solar cell production. We will discuss problems arising in thin-film cells without reference to the production technique except for those cases specific to a particular deposition technology.

8.2   MATERIAL AND STRUCTURAL CHARACTERISTICS AFFECTING CELL PERFORMANCE

1. Stoichiometry

As a general rule thin films are more susceptible to variations in stoichiometry than bulk materials both during and after formation. Effects at the surface such as chemical reaction, oxidation, or the presence of surface states may affect a substantial proportion of the active region of a

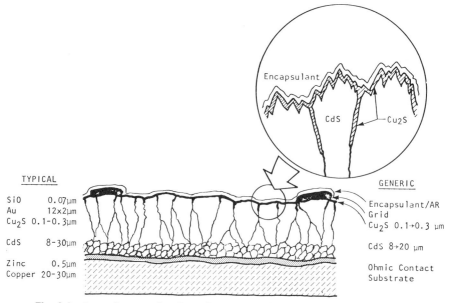

Fig. 8.4 Approximate scale schematic representation of the structure of the thin-film CdS–Cu$_2$S solar cell.

thin-film cell. Deposition and the postdeposition environment must accordingly be closely controlled. In many cases stoichiometry controls the resistivity [20] of the component materials. As a first-order effect stoichiometry may determine carrier density. The CdS–Cu_2S cell provides examples of stoichiometry control both during film formation and subsequent exposure to the atmosphere. In the CdS the sulfur vacancy concentration is determined by the mode of formation and deposition conditions, yielding an *n*-type semiconductor of very widely varying resistivity. In the Cu_2S layer, the presence of copper vacancies produces a *p*-type material. This layer, generally only on the order of 1000 Å thick, is particularly sensitive to surface changes. The most important effect is the formation of a surface copper oxide which can change the carrier density from less than 10^{17} to the order of 10^{21} holes/cm^3. (See Fig. 8.4.) One or two monolayers of copper oxide may be sufficient to convert the Cu_2S from a nondegenerate to a highly degenerate semiconductor with major effects on cell performance. Similar effects with significant practical consequences are certain to be present in other thin-film solar cells. The factors determining stoichiometry are dependent upon the means of preparation of the film [21], i.e., substrate and source temperatures for evaporation processes, sticking coefficients, arrival rates, and nucleation sites.

2. Grain Size

The dominating structural difference between thick single crystals and most thin-film cells is the presence of grain boundaries. These have an electronic effect, because they represent a generally potent source of recombination. In the extreme, the ability of a cell to generate measurable short-circuit current can be eliminated. Under less extreme circumstances the effect of grain boundaries will be determined by the grain size, more specifically the ratio of grain size to film thickness, and the grain boundary recombination velocity.

In Fig. 8.5 we have illustrated a cylindrical model for grains. With a simple geometric analysis the regions in which generated carriers will be lost, assuming totally effective grain boundary recombination, are as shown [22]. In Fig. 8.6 we have illustrated the calculated results for this polycrystalline model as a function of the αd product and the grain radius to film thickness (r/d) ratio for two device configurations, front wall and back wall [23]. These results show that the grain radius must be large compared with the absorbing material thickness if an appreciable fraction of the short-circuit current potentially available is to be realized in practice.

Grain size related structural effects can also occur, as illustrated in the Cu_2S–CdS heterojunction. During the chemical formation of the Cu_2S

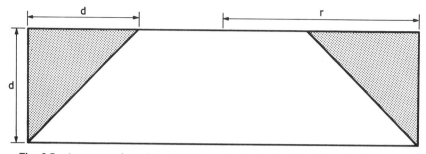

Fig. 8.5 A cross section of a cylindrical grain illustrating (shaded regions) those areas which will not contribute to current generation because of the loss of carriers to the grain boundaries when light is incident through the top surface and the collection junction is the bottom surface.

layer it is observed that the reaction takes place preferentially at the CdS grain boundaries. In consequence the final junction area A_j is considerably larger than the normal projection of the CdS layer $A_\perp$. This effect multiplies j_{03} by $A_j/A_\perp$ and reduces the open-circuit voltage of the cell as illustrated in Fig. 8.7. The calculated variation of the open-circuit voltage of a cell is given as a function of the Cu_2S equivalent thickness which is related to the degree of Cu_2S penetration down grain boundaries [22]. The loss in open-circuit voltage is minimized by achieving large r/d ratios.

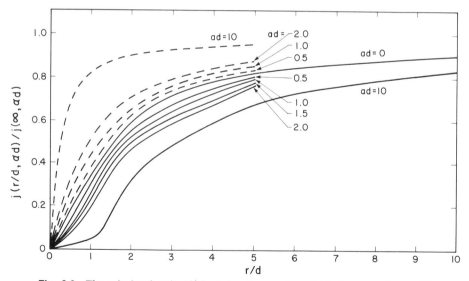

Fig. 8.6 The calculated ratio of short-circuit current available from a polycrystalline solar cell to that from a single-crystal cell as a function of grain radius r, collecting layer thickness d, absorption coefficient α, and whether the light passes through the absorbing layer before reaching the junction (front wall, solid line) or after (back wall, broken line), assuming total loss of carriers that reach a grain boundary.

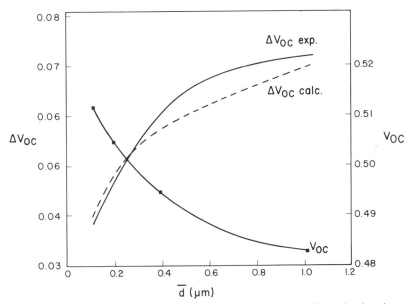

Fig. 8.7 Calculated (–––) and experimental (———) results illustrating loss in open-circuit voltage (■) due to the formation of Cu_2S down grain boundaries in the CdS–Cu_2S solar cell.

Another grain boundary effect which can reduce open-circuit voltage is recombination at grain boundaries in the space charge region [24, 25]. This increases j_{02} by reducing τ in the vicinity of the grain boundary. This effect is also reduced by having a large r/d ratio. Fraas has recently published a review of many grain boundary effects [26].

3. Surface Properties

Surfaces are of primary concern because of the presence of surface states and the associated effects on device behavior. Based on the nature of the surface states, the energy bands can be bent so that minority carriers are either repelled or attracted to the surface. In the event that minority carriers are attracted to the surface, recombination is promoted, substantially reducing the short-circuit current obtainable from the cell. The effect on short-circuit current of the effective surface recombination rate can be seen in Fig. 8.8. In order for a solar cell to have good conversion efficiency the effective surface recombination velocity must be small. This is generally described as surface passivation which can be achieved in two fundamentally different ways. An appropriate doping profile will repel minority carriers from the surface, preventing surface recombination. Alternatively a second layer can be created on the active absorbing material so that the

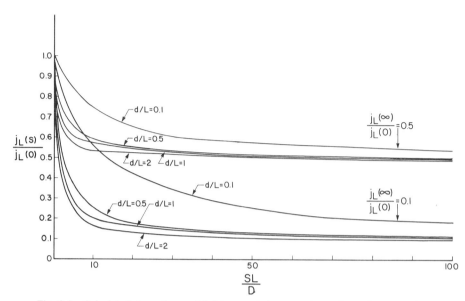

Fig. 8.8 Calculated dependence of light-generated current j_L on the surface recombination velocity S and d/L ratio. (See also Fig. 8.12.)

surface states are neutralized, preventing them from acting as recombination centers.

In addition to losses in current, surface recombination (and also back contact or grid contact recombination) reduces the open-circuit voltage by increasing j_{01}. Because the semiconductors have finite thicknesses, the change in boundary conditions produced by the presence of surface recombination modifies the expression for j_{01}; e.g., for a p layer of thickness d, diffusion length L_n, and diffusion constant D_n, the term $D_n n_{0p}/L_n$ is multiplied by the factor [2]

$$\frac{1 + (D_n/SL_n)\tanh(d/L_n)}{D_n/SL_n + \tanh(d/L_n)} \qquad (8.9)$$

where S is the surface recombination velocity. For large SL_n values and $d \ll L_n$, this factor approaches L_n/d. Each order-of-magnitude increase in j_0 corresponds to a loss in V_{oc} of ~ 60 mV.

4. Grain Boundary Properties

Recombination at grain boundaries has been shown to be a potential source of serious carrier losses with corresponding reductions in short-circuit current. Junction area effects related to grain boundaries may also reduce the achievable V_{oc}. Grain boundaries may also result in enhanced

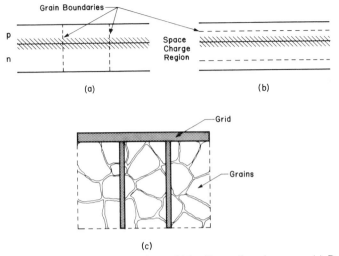

Fig. 8.9 Grain boundary configurations which affect cell performance. (a) Boundaries perpendicular to junction lower j_{sc} and V_{oc}; (b) boundary parallel to junction lowers j_{sc}; (c) isolated grain lowers j_{sc}.

diffusion of dopants, as has been observed in polycrystalline silicon. During a conventional diffusion anneal enhanced dopant penetration occurs down the grain boundaries. In extreme cases penetration right through the base layer can completely short out the device. Under less severe conditions a nonplanar junction results. The grain boundaries can, in some cases, act beneficially by gettering impurities from the bulk, and it has been reported that appropriate doping results in band bending so as to eliminate recombination at the boundaries [1, p. 108].

Grain boundaries may also affect the transport of majority carriers. Scattering effects or the existence of energy barriers can determine the effective carrier mobility and the resistivity of polycrystalline layers [27]. Macroscopic effects will then be apparent in increased series resistance and lower fill factors. Isolated grains smaller than the grid spacing will cause a loss in j_{sc}.

We have used a simple cylindrical grain model to discuss the types of problems that can occur in polycrystalline cells. The cylindrical grain is admittedly a major simplification but is probably adequate for treating films showing columnar growth. Other configurations of grain boundaries occur and may cause more severe problems. For example, a grain boundary parallel to the junction plane may restrict useful current generation to the region between the grain boundary and the junction. A number of grain boundary configurations are illustrated in Fig. 8.9.

5. Bulk Lifetime

Deposition conditions, stoichiometry, and localized energy levels associated with dislocations or impurities may all influence minority carrier lifetimes. This parameter directly influences overall cell performance through either the short-circuit current or the open-circuit voltage or both. In a homojunction the lifetime τ influences both short-circuit current and open-circuit voltage. Whereas for at least some heterojunctions τ primarily affects the short-circuit current. In thick single-crystal solar cells the bulk lifetime must of necessity be long to achieve adequate quantum efficiency. The situation in thin-film solar cells is fundamentally different. Highly absorbent materials perhaps coupled with cell designs giving multiple light passes result in thin layers being adequate for effective absorption and minority carrier generation. As a consequence, although the bulk lifetimes must still exceed some minimum value, these will be orders of magnitude smaller than those needed for thick single-crystal cells as exemplified by silicon. Purity and perfection limits are correspondingly relaxed.

6. Energy-Band Profiles

Whether the junction results from conventional diffusion doping, heterojunction formation, or by controlling stoichiometry, spatially varying energy levels can be formed. This is generally referred to as band bending which acts in all respects like an internal electric field. Such a field can produce several beneficial effects. Minority carriers created within the field can be swept more rapidly toward the interface, with a resulting increase in short-circuit current. Surface passivation caused by suitable fields near the surface has already been discussed. The doping density may also directly influence carrier behavior. The mobility in polycrystals, for low doping concentration, is generally controlled by the same phenomena as in single-crystal material, e.g., lattice scattering by either acoustic phonons or optical phonons and piezoelectric scattering. In highly doped material, the mobility will be controlled by either neutral or charged-impurity scattering, resulting in reduced mobility and shorter diffusion lengths. Further effects are observed at very high doping densities. Lifetimes may be reduced by Auger mechansims [28], and shrinkage of the band gap [29] may cause detectable losses in open-circuit voltage.

7. Impurities

We will distinguish between impurities and the intentional dopants which give n- or p-type character to the materials. Impurities will generally have a negative effect on the cell performance and may derive from the starting materials or become incorporated during cell fabrication. A range of deleterious effects can be traced to impurities such as deep levels which reduce carrier lifetime and diffusion length. Impurities may also lower mobility or create traps and charge centers which can affect the current–voltage behavior of the cell [16].

8. Dislocations

Dislocations can act directly as trapping and recombination centers or, by attracting impurities, become preferred sites for recombination, etc. In either case carrier lifetime and mobility may be adversely affected, and dislocation populations must be limited to achieve high conversion efficiency. Specific dislocation structures can be expected in most heterojunctions with significant effects on the interface recombination kinetics.

9. Interface States

The population of interface states in heterojunctions can be expected to depend on the formation technique, the component materials, and the crystallography of the junction. Even in ideal cases, the mismatch in lattice constants will give rise to misfit dislocations with associated dangling bonds and interface states (see Fig. 8.10). Interface states may dominate the dark current–voltage curve and can also affect the short-circuit current of the cell. In the CdS–Cu$_2$S solar cell, recombination at interface states acts as a gate controlling carrier flow, hence the short-circuit current [30]. The relation given below shows that the interface recombination velocity is an extremely important parameter in CdS–Cu$_2$S and heterojunction cells in general.

In the CdS–Cu$_2$S solar cell interface states can trap electrons and provide the path by which they recombine with holes in the Cu$_2$S layer. Equation (8.4) describes the dark current, and the light-generated current is

$$j_L(V) = j_{L0}\, \frac{\mu_2 F_2(V)}{S_I + \mu_2 F_2(V)} \tag{8.10}$$

Fig. 8.10 Edge dislocation at interface as source of interface states.

The junction field F_2 tends to sweep electrons that have crossed into the CdS away from the interface, thus preventing their capture by the interface states. The electric field, hence j_L, is voltage-dependent such that j_L is reduced for $V > 0$. This decline in j_L with voltage to a degree determined by the voltage dependence of F_2 will reduce both the fill factor and V_{oc}.

The interface states in the CdS–Cu$_2$S cell have been shown to control all three major cell parameters, short-circuit current, fill factor, and open-circuit voltage. It is likely that this will be generally true for heterojunctions, hence control of interface state behavior is the key to producing successful thin-film solar cells from dissimilar materials.

The interface state density N_I is proportional to the lattice mismatch and, for rectangular lattices with $a_i > b_i$,

$$N_I = \frac{a_1 - b_1}{a_1 b_1 b_2} + \frac{a_2 - b_2}{a_2 b_1 b_2} \tag{8.11}$$

The misfit dislocation array at the interface depends not only upon the lattice mismatch but also upon the thickness of the layers. The initial layers of atoms deposited or grown on a bulk substrate will be strained to match the lattice constant of the substrate, but at a critical thickness, which depends upon the lattice mismatch and elastic constants of the materials, it becomes energetically favorable for misfit dislocations to form [31].

8.3 OPTICAL PROPERTIES

The optical properties of thin films should not differ significantly from those of the bulk single crystal. The optical band gap should be virtually the same except for strain effects. Absorption coefficients are an intrinsic property of the material and should not be significantly affected by going to thin films. The experimental measurement of the optical constants of thin films is beset with difficulties and, as indicated in Chapter 4, considerable care must be taken to generate reliable results. Problems arise because of internal boundaries, mainly grain boundaries, which scatter and diffuse the light. In a traditional transmission experiment with normal incidence, scattering within the film may be measured as an apparent absorption. This spurious absorption effect due to scattering of the light out of the primary beam can be minimized but not eliminated by using a 4π integrating reflectance sphere [32]. Scattering at grain boundaries can result in some light approaching the exit surface at an angle larger than the critical angle for internal reflection. As a result, light is trapped within the film, giving an apparent absorption above that for a single-crystal film of the same thickness.

The surface region represents a large fraction of the volume of a thin film, and extra absorption at wavelengths longer than the bulk band gap can occur as a result of band bending. This effect can arise in both single and polycrystalline films. Band bending or gap shrinkage can also occur near grain boundaries, creating a further mechanism for enhanced absorption in a polycrystalline film.

8.4 ELECTRICAL PROPERTIES

1. Diffusion Length

The diffusion length is the parameter most directly controlling the short-circuit current of the cell, and it can also influence the open-circuit voltage if the diode current is controlled by the diffusion length of the material [see Eqs. (8.2) and (8.9)]. The diffusion length L is given by $L = (kT\tau\mu/q)^{1/2}$, showing that mobility and recombination lifetime fundamentally control L. Lifetime controlling mechanisms such as grain boundary, interface, and surface recombination have already been discussed, and each will cause a change in the effective diffusion length. It is generally possible to separate the effects of surface and grain boundary recombination and bulk lifetime effects by careful measurement. Quite generally diffusion length should be as long as possible and certainly equal to the thickness of the absorbing layer. The dependence of short-circuit current on diffusion length is shown in Fig. 8.11 [33]. The deleterious effect of a short diffusion length can be overcome to an extent by creating a built-in field with an appropriate doping profile. In the case of a Mott-type Schottky diode, a field exists as a consequence of the diffusion voltage. Compensating the semiconductor near the junction will increase the field width and accordingly enlarge the region in which the diffusion length exceeds that found in the field-free region. As the equations below illustrate, the presence of a field of sufficient magnitude can considerably enhance the effective diffusion length, hence the short-circuit current. This is particularly important in materials like amorphous silicon in which the intrinsic diffusion length is very small.

The effective diffusion length in the presence of an electric field is increased if the carrier is proceeding with the field or reduced for an opposing field. The field affected diffusion length L is given by [34].

$$L_{1,2}^{-1} = (2L)^{-1}\left\{\left[(F/F_c)^2 + 4\right]^{1/2} \pm F/F_c\right\} \qquad (8.12)$$

where L is the field-free diffusion length and $F_c = kT/qL$.

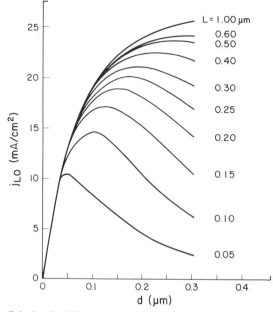

Fig. 8.11 Calculated AM1 current densities for a CdS–Cu$_2$S solar cell as a function of Cu$_2$S thickness and diffusion length assuming reflection from back contact.

2. Mobility

Lattice scattering generally dominates the mobility in pure materials, but in thin films grain boundary scattering can significantly affect the observed mobility. As illustrated in Chapter 3, the mobility measured across or parallel to grain boundaries can be significantly different. Thus mobility can be both an important parameter controlling cell performance and very difficult to measure in real devices. Further complications arise as a result of impurity concentrations and gradients which also affect mobility and hence diffusion lengths.

It is worth noting that not one but several mobilities control thin-film solar cells. The minority carrier mobility controls the diffusion length to the first order. If we are dealing with columnar crystallites, then the mobility in question is that for flow parallel to the grain boundaries. Majority carrier mobilities may control the series resistance of the cell. As a specific example, consider the front wall (illuminated through the Cu$_2$S) CdS–Cu$_2$S solar cell in which there are two major contributions to series resistance. The p-type Cu$_2$S layer contributes to R_s through its sheet resistance controlled by the lateral flow of holes across grain boundaries to the grid contacts. For parallel line grids the relation is

$$AR_{s1} = (\rho/d_1)(s^2/12) = s^2/12d_1 q p \mu_p \qquad (8.13)$$

where A is the area of the cell and s the spacing between the grid fingers. The hole mobility μ_p depends upon the carrier concentration p, through impurity scattering from ionized acceptor Cu vacancies ($[Cu]_{vac} = p$). In addition the barrier to current flow through the grain boundaries will also be dependent on carrier concentration.

The contribution to the series resistance from the CdS layer is given by

$$A R_{s_2} = \rho d_2 = d_2 / q n \mu_n \qquad (8.14)$$

where d_2 is the width of the compensated region of the CdS layer. Here the controlling mobility is that for flow parallel to the columnar grains. The same mobility enters into the interface collection factor [see Eq. (8.10)].

3. Seebeck Coefficient

The thermoelectric power or Seebeck coefficient can be used to obtain a measure of the position of the Fermi level, hence a measure of the carrier concentration in the materials [35–37]. However, since the Seebeck coefficient depends on the scattering mechanisms present in the material, the presence of dislocations, grain boundaries, interfaces, and other scattering mechanisms may make the unambiguous interpretation of the Seebeck measurements difficult.

4. Defects

We have already mentioned the fact that defects can act as trapping and recombination centers. The electrical effect of these would be in the changes in field profiles if the defects can act as hole or electron sites of storage. In the case of the CdS–Cu$_2$S cell, the diffusion of copper into the CdS layer produces a high density of defects which are acceptors in the dark but which are easily emptied under appropriate wavelength light. In the light, the space charge region in the CdS shrinks dramatically, increasing the photocapacitance by as much as a factor of 100 [30]. Similar effects have also been reported in amorphous silicon [38]. Such shrinkage of the space charge region under illumination strongly affects the junction field and may alter the interface collection factor and short-circuit current [Eq. (8.10)]. The spectrally sensitive collection efficiency of the CdS–Cu$_2$S cell is due to this effect. Defects can also act directly to enhance recombination and reduce the short-circuit current.

8.5 OPTIMAL CELL DESIGN CONSIDERATIONS

1. Electron Affinity Match

Interface states almost certainly exist in all heterojunctions, in which case the electron affinity difference between the two materials determines the magnitude of the barrier which the minority and majority carriers see as they cross the interface. In the case of the CdS–Cu$_2$S solar cell, the electron affinity mismatch is on the order of 0.2 eV in the direction which reduces the open-circuit voltage of the cell (see Fig. 8.2). To change the electron affinity mismatch it is necessary either to alloy the material significantly or to change totally one of the junction components. For example, the substitution of 20%–30% zinc for cadmium in CdS produces an electron affinity match and results in a cell with a higher open-circuit voltage than is possible with pure CdS.

2. Lattice Constant Match

In heterojunction solar cells, the mismatch in lattice constant between the two materials determines the density of interface states [Eq. (8.11)], which may in turn place limits on the short-circuit current open-circuit voltage and fill factor which can be achieved. A primary selection criterion is therefore to pick materials which have lattice constants as well matched as possible.

3. Grid Contact

The grid contact on a solar cell is an important efficiency-controlling design parameter. It is obvious that transmission through the grid should be maximized to the degree compatible with other requirements. The grid material itself must make ohmic contact to the contiguous semiconductor. The grid design influences the series resistance of the cell because of the lateral current flow to the grid wires [Eq. (8.13)]. If a transparent conducting material of sufficient transmission and conductivity were available, it would in many ways be preferable to the conventional grid structure. No such material exists at present, and a current-collecting grid represents an available technology for applying contacts to a solar cell. Equation (8.15) gives the optimum spacing s between grid lines in terms of the sheet resistance of the top layer ρ_1/d_1:

$$s \simeq \left[\frac{6V_{oc}d_1\delta}{Cj_{sc}\rho_1} \left(FF_0 - C\frac{j_{sc}}{V_{oc}} R_{s2}A \right) \right]^{1/3} \qquad (8.15)$$

where FF_0 is the theoretical fill factor for a given open-circuit voltage V_{oc}, δ the width of the grid line, C a slowly varying function of V_{oc} ($C = 0.9$ for $V_{oc} = 0.5$), R_{s2} the contribution to the series resistance not due to the top layer, and A the area of the cell.

4. Layer Thicknesses

The optimum thicknesses of the layers comprising a thin-film cell are determined by a number of boundary conditions. For the absorber layer,

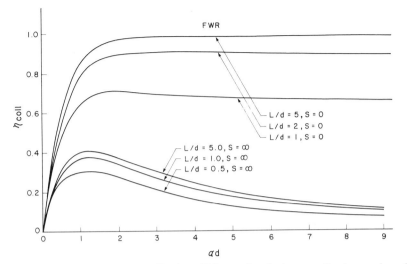

Fig. 8.12 Calculated current-collection efficiency of a single generating layer solar cell as a function of αd product, L/d ratio, and surface recombination velocity, assuming front-wall mode of operation and reflection from the back contact.

the first requirement is that the product of absorbance and thickness αd be sufficient to ensure that virtually all the useful solar spectrum is absorbed and electron–hole pairs generated. This may be achieved in a single pass, or in multipasses as a result of reflection from the back-contact and light-trapping effects. While satisfying the absorption requirement the thickness must also be such that the minority carrier diffusion length L is equal to or greater than the thickness of the absorbing layer. Some quantitative results of a computation of quantum efficiency, as influenced by thickness, diffusion length, and absorption are shown in Fig. 8.12. There is a third requirement influencing the thickness of the semiconductor layers in a cell arising from sheet resistance effects. This may be strongly coupled to grid or substrate design, but the overall requirement is that the cell fill factor is not unacceptably reduced by a series resistance term.

In polycrystalline cells there is a potential problem from shunting paths arising from enhanced grain boundary penetration of dopants or impurities. An allied problem can arise in the CdS–Cu$_2$S cell if the grain boundary penetration of Cu$_2$S occurs through the full thickness of the CdS layer.

5. Surface Passivation

As indicated in Fig. 8.8, the surface recombination velocity can be a controlling factor in the short-circuit current, hence in the efficiency of a solar cell. For an efficient thin-film cell it is essential that effective surface passivation prevent carrier loss by recombination at external or internal surfaces. It must be concluded that the Cu_2S layer on a CdS solar cell is self-passivating, most probably by the formation of an oxide layer. Gallium arsenide does not self-passivate, and a layer of GaAlAs must be applied to reduce surface recombination to acceptable levels. In other materials it is found that chemical polishing will reduce the surface recombination velocity. A somewhat different approach is to produce a doping profile, causing an internal electric field opposing minority carrier motion toward the surface. In single-crystal silicon cells, this is known as a back-surface field and is generated by a p^+ (heavily p-type) region between the normal p-type base and the back contact [1].

6. Photon Economy

It is essential to minimize the amount of light reflected from a solar cell, and two classes of techniques are presently in use. The first relies on texturing the entrance surface so that reflected light intercepts another entrance surface. Such texturing is created on the Cu_2S–CdS solar cell by chemical etching of the CdS layer prior to the Cu_2S formation. This produces a quite dramatic reduction in reflection at all wavelengths, as shown in Fig. 8.13 [39]. Very similar surface texturing has been used for single-crystal silicon cells. The second approach is to use refractive index matching and interference thickness antireflection coatings. For ideal planar surfaces zero reflection can be achieved at a specific wavelength λ

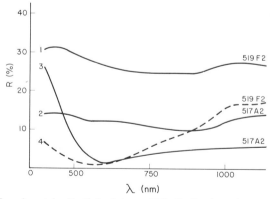

Fig. 8.13 Experimental reflectivity data on CdS–Cu_2S solar cells illustrating the effects of antireflection (AR) coatings (SiO), etching of the surface, and method of formation of the Cu_2S layer. (1) Unetched, no AR, solid-state reacted Cu_2S; (2) Unetched, no AR, solution-grown Cu_2S; (3) Same sample as in (1) with 751 Å of SiO_x; (4) Same sample as in (2) with 751 Å of SiO_x.

with a coating of thickness d if $d = \lambda/4n_2$. Here n_2 is the refractive index of the coating and n_1 that of the semiconductor, with $n_2 = (n_1)^{1/2}$. The reflection shows a relative broad minimum as a function of wavelength, hence low total reflectivity over the active band of the solar spectrum can be achieved. Typical antireflection materials are SiO, Nb_2O_5, Ta_2O_5, Si_3N_4, TiO_2, and Sb_2O_3. These and a number of others with indexes of refraction in the neighborhood of 2 have been used successfully to cut down the first surface reflection. The use of a two-layer antireflection coating can be even more effective by broadening the wavelength range of minimum reflectance.

7. Substrate Properties

For many solar cells an important substrate characteristic is a lattice constant matched to the material deposited. This will promote an epitaxial growth mode and good adherence to the surface. Good adherence is also assisted by matching the thermal expansion constants of the substrate and the material being deposited. Deposition generally takes place at temperatures well above room temperature, and differences in thermal expansion create possibly destructive stresses on cooling. The optical properties of the substrate, specifically the reflection coefficient, may also strongly influence the cell conversion efficiency. There is good reason to keep the absorbing layer as thin as possible to minimize carrier recombination and to achieve good carrier collection in spite of short diffusion lengths. If the light only makes one pass through the absorber, utilization of the incident photons may be very incomplete. However, if the substrate is highly reflecting, light not absorbed on the first pass can be utilized on a second or subsequent pass. For multiple reflections, major photon loss could occur at the back contact, hence its reflectivity is extremely important. A criterion for high reflectivity by a metal is that the real part of the index of refraction be as small as possible. In this case the reflectivity given by the relation

$$r = \left[(n_1 - n_2)^2 + (k_1 - k_2)^2 \right] / \left[(n_1 + n_2)^2 + (k_1 + k_2)^2 \right]$$

will be nearly unity. The metals which satisfy the criterion that the real part of the index of refraction be nearly zero in the optical range are silver and copper.

8.6 RECENT RESULTS

1. Performance of Thin-Film Cells

A wide variety of materials have been used as the basis of thin-film solar cells. Table 8.1 shows the performance achieved with a number of

TABLE 8.1

Representative Thin-Film Solar Cell Characteristics

Cell	J_{sc} (mA/cm^2)	V_{oc} (V)	FF	η (%)	Lighting	Reference
CdS–Cu$_2$S	21.8	0.52	0.71	9.15	87.9 mW/cm^{2a}	[10, p. 419]
InP–CdS	13.5	0.46	0.68	5.7	AM2	[14, p. 483]
a-Si–Pt	12.0	0.80	0.58	5.5	60 mW/cm^2	[14, p. 449]
n-CdSe–p-ZnTe	1.89	0.56	0.48	0.60	85 mW/cm^2	[14, p. 487]
n-CdSe–p-CdTe	0.75	0.61	0.30	0.16	85 mW/cm^2	[14, p. 487]
n-CdS–p-CdTe	13.9	0.59	0.55	5.2	85 mW/cm^2	[14, p. 487]
p-Cu$_2$Te–nCdTe	13.0	0.59	0.63	4.8	100 mW/cm^2	[5, p. 33]
CdS–CuInSe$_2$	21.0	0.46	0.54	5.2	100 mW/cm^2	[14, p. 496]
Poly Si	23.5	0.57	0.72	9.5	AM1	[10, p. 1106]
Cu$_2$O–Cu	7	0.35	0.45	1.1	AM1	[10, p. 174]
GaAs–Au	12.5	0.46	0.53	2.25	AM0	[10, p. 960]

^{a}Natural sunlight. All others are simulated sunlight.

thin-film cells. These figures should be interpreted with some caution and in general should not be taken as indicative of the achievable efficiencies. Poor efficiency can result from various causes many of which are of an engineering nature. Included in this category are gridding and antireflection techniques which may be used to increase substantially the measured efficiency. On the other hand, unjustifiably high values have been reported by using only the active area (omitting the area covered by grids), correcting for all reflection losses, and using best values from very small devices. The high efficiencies reported will not be attainable in a practical device because some loss due to reflection and gridding losses in large-area devices are unavoidable. Furthermore, in large-area polycrystalline devices, material nonuniformity may reduce the average efficiency to well below the best small-area value.

The theoretical efficiency for a given solar cell can be calculated and then, by analyzing the unavoidable losses, one can compute the attainable efficiency. Although single-crystal cells have achieved practical efficiencies much closer to the theoretical limit than polycrystalline thin-film cells, there seems to be no fundamental reason why thin-film cells should not also reach $\sim 80\%$–90% of the no-loss value.

2. Loss Analysis

To determime the attainable efficiency for a given cell design the theoretical (no-loss) efficiency is first computed for a given spectrum of light. The no-loss efficiency is obtained by assuming (1) that all incident photons of more than band-gap energy contribute one electron to the short-circuit current, (2) that the open-circuit voltage is limited only by the

TABLE 8.2

Factors Determining J_{sc} in a Thin-Film $CdS-Cu_2S$ Solar Cell and Typical Values in Percent

Absorption reflection	Grid shading	Surface loss	Bulk recombination	Interface loss	Grain boundary	CdS and back-contact light loss	Total losses
5	5	1	5	5	1	10	28

smallest band gap in the junction, and (3) that the fill factor is not reduced by either series or shunt resistance effects.

Achievable short-circuit currents may be reduced by both optical loss of photons and electrical loss of carriers. Table 8.2 lists the loss mechanisms to be considered for any polycrystalline cell [17]. The attainable short-circuit current is the no-loss current multiplied by the product of one minus the factional loss due to each factor, i.e.,

$$j_{sc} = j_{sc0} \prod_i (1 - l_i)$$

Table 8.2 illustrates the calculation for a particular CdS–Cu$_2$S cell design and shows how even quite small individual loss mechanisms combine to give an achievable performance substantially below the ideal no-loss value.

A similar computation can be carried out for the open-circuit voltage. Table 8.3 illustrates this approach for the CdS–Cu$_2$S system [17]. In this case the losses are simply additive.

The fill factor depends upon V_{oc}/kT and the series and shunt resistances in a nonsimple manner. However, for good cells the shunt effects are negligible, and the series resistance dependence can be linearized to give a simple expression valid for cells operated at $\sim$AM1. The losses in fill factor due to the sheet resistance of the top layer, the grid metal, and the base material can then be calculated easily. There are losses due to other factors which may occur in specific cells. For example j_L may be voltage-dependent. In CdS–Cu$_2$S cells this may be due to the interface recombination mechanism [Eq. (8.10)], and, in amorphous Si Schottky cells, the field dependence of the diffusion length [Eq. (8.12)] may have the same effect. Table 8.4 shows the major mechanisms causing a reduction in achievable fill factor for a particular CdS–Cu$_2$S cell design.

TABLE 8.3

Typical Values of Quantities That Determine V_{oc} in the $CdS-Cu_2S$ Solar Cell

$E_{g1} - \Delta X$	$+ kT \ln J_{sc}$	$- kT \ln q N_{c2} S_I$	$- kT \ln A_j/A_\perp$	Calculated V_{oc}
1.0	$- 0.1$	$- 0.33$	$- 0.06$	0.51

TABLE 8.4

Typical Values of Terms That Determine the Fill Factor in the CdS–Cu$_2$S Solar Cell

Series resistance		Voltage dependence of J_L	Calculated *FF*
Cu$_2$S	CdS		
− 0.04		− 0.05	0.71

8.7 CONCLUSIONS

The treatment of thin-film photovoltaic devices presented in this chapter has avoided detailed discussion of film growth techniques and specific material properties of films, which are covered in other chapters. However, it is ultimately the details of the growth technique and the control variables of source and substrate temperature, growth rate, and substrate properties which determine the performance of the devices through key material properties such as diffusion length, grain size, grain barrier height, mobility, surface recombination velocity, and optical constants. The variation in cell properties resulting from different preparation conditions is almost infinite. The range of results obtained in the CdS–Cu$_2$S system [40] illustrates this variability. The properties of the thin-film cells can, however, be successfully modeled [17, 41].

REFERENCES

1. H. J. Hovel, "Semiconductors and Semimetals," Vol. 11, Solar Cells. Academic Press, New York, 1975.
2. A. Rothwarf and K. W. Böer, *Prog. Solid-State Chem.* **10**, 71–102 (1975).
3. K. W. Böer and A. Rothwarf, *Ann. Rev. Mater. Sci.* **6**, 303–333 (1976).
4. Conference Record, *IEEE Photovoltaic Specialists Conf., 7th, Pasadena, California* (1968).
5. Conference Record, *IEEE Photovoltaic Specialists Conf., 8th, Seattle, Washington*, Catalog No. 70C 32 ED, Inst. of Electrical and Electronic Engineers, 345 East 47th Street, New York, 10017 (1970).
6. Conference Record, *IEEE Photovoltaic Specialists Conf., 9th, Silver Springs, Maryland* (1972).
7. Conference Record, *IEEE Photovoltaic Specialists Conf., 10th, Palo Alto, California* Cat. No. 73CHO 801-ED (1973).
8. Conference Record, *IEEE Photovoltaic Specialists Conf., 11th, Scottsdale, Arizona* Cat. No. 75 CHO948-OED (1975).
9. Conference Record, *IEEE Photovoltaic Specialists Conf., 12th, Baton Rouge, Louisiana* Cat. No. 76 CH 1142-ED (1976).
10. Conference Record, *IEEE Photovoltaic Specialists Conf., 13th, Washington, D. C.* 78 CH1319-3ED (1978).

11. C. Backus (ed.), "Solar Cells." IEEE Press, New York, 1976.
12. *Photovoltaic Solar Energy Conf., Luxembourg, September 27–30, 1977.* Reidel, Boston, Massachusetts, 1978.
13. *Int. Conf. Solar Elec., Toulouse.* Centre National D'Etudes Spatiales, Toulouse, France, 1976.
14. *IEEE Trans. Electron Devices* **ED-24** (1977), Special Issue on Photovoltaic Devices.
15. J. W. Matthews (ed.), "Epitaxial Growth," Parts A and B. Academic Press, New York, 1975.
16. A. Rothwarf, Conference Record, *IEEE Photovoltaic Specialists Conf., 13th, Washington, D. C.* No. 78 CH1319-3ED, p. 1312 (1978).
17. A. Rothwarf and A. M. Barnett, *IEEE Trans. Electron Devices* **ED-24**, p. 381 (1977).
18. A. Rothwarf, International Workshop on Cadmium Sulfide Solar Cells and Other Abrupt Heterojunctions, University of Delaware, May 1975, NSF-RANN AER 75-15858, p. 9.
19. J. A. Amick, G. L. Schnable, and J. L. Vossen, *J. Vac. Sci. Technol.* **14**, 1053 (1977).
20. B. Baron, A. W. Catalano, and E. A. Fagen, Conference Record, *IEEE Photovoltaic Specialists Conf., 13th, Washington, D. C.* No. 78 CH1319-3ED, p. 406 (1978).
21. L. L. Chang and R. Ludeke, *in* "Epitaxial Growth" (J. W. Matthews, ed.), Parts A and B, p. 37. Academic Press, New York, 1975.
22. A. Rothwarf, Conference Record, *IEEE Photovoltaic Specialists Conf., 12th, Baton Rouge, Louisiana* Cat. No. 76 CH1142-ED, p. 488 (1976).
23. A. Rothwarf, L. Burton, H. Hadley, and G. M. Storti, Conference Record, *IEEE Photovoltaic Specialists Conf., 11th, Scottsdale, Arizona*, p. 476 (1975).
24. C. T. Sah, Conference Record, *IEEE Photovoltaic Specialists Conf., 12th, Baton Rouge, Louisiana* Cat. No. 76 CH1142-ED, p. 93 (1976).
25. H. C. Card and E. S. Yang, *IEEE Trans. Electron Devices* **ED-24**, 397 (1977).
26. L. M. Fraas, *J. Appl. Phys.* **49**, 871 (1978).
27. L. L. Kazmerski, W. B. Berry, and C. W. Allen, *J. Appl. Phys.* **43**, 3516, 3521 (1972).
28. D. Redfield, *Appl. Phys. Lett.* **33**, 531 (1978).
29. F. A. Lindholm and C. T. Sah, *IEEE Trans. Electron. Devices* **ED-24**, 299 (1977).
30. A. Rothwarf, J. Phillips, and N. C. Wyeth, Conference Record, *IEEE Photovoltaic Specialists Conf., Washington, D. C.* No. 78 CH1319-3ED, p. 399 (1978).
31. C. A. B. Ball and C. Laird, *Thin Solid Films* **41**, 9 (1977).
32. E. A. Fagen *J. Appl. Phys.* **50**, 6505 (1979).
33. A. Rothwarf, *Int. Conf. Solar Electricity, Toulouse* p. 273, Centre National D'Etudes Spatiales, Toulouse, France, 1976.
34. R. A. Smith, "Semiconductors." Cambridge Univ. Press, London and New York, 1968.
35. Y. Y. Ma and R. H. Bube, *J. Electrochem. Soc.* **124**, 1430 (1977).
36. F. Guastavino, H. Luquet, and J. Bougnot, *Int. Cong. "The Sun in the Service of Mankind," Paris* p. 189. Centre National D'Etudes Spatiales Bretigny-Sur-Orge, France, 1973.
37. G. Z. Idrichan and G. P. Sorokin, *Inorg. Mater.* **11**, 1449 (1975).
38. C. R. Wronski, *IEEE Trans. Electron Devices* **ED-24**, 351 (1977).
39. J. A. Bragagnolo, Conference Record, *IEEE Photovoltaic Specialist Conf., Washington, D. C.* No. 78 CH1319-3ED p. 412 (June 1978).
40. A. G. Stanley, *Appl. Solid State Sci.* **5**, 251 (1975).
41. L. L. Kazmerski, P. J. Ireland, F. R. White, R. B. Cooper, Conference Record, *IEEE Photovoltaic Specialists Conf., Washington, D. C.* No. 78 CH1319-3ED, p. 184 (June 1978).

9 Applications of Passive Thin Films

PATRICK J. CALL

Solar Energy Research Institute
Golden, Colorado

9.1 INTRODUCTION

A remarkable number of applications exists for thin-film technology across the spectrum of solar energy conversion devices. Passive films, defined to exclude applications such as thermoelectrics and photovoltaics

257

POLYCRYSTALLINE AND AMORPHOUS
THIN FILMS AND DEVICES

where the film itself is the primary transducer or conversion element, are critical to nearly every solar technology. The category of passive thin films as defined in this chapter includes optical films, protective coatings, high and low-energy surfaces, and selective membranes.

Optical films for solar energy conversion systems include absorber, reflector, and enhanced-transmission (antireflection) materials, heat mirrors, and transparent conducting electrodes (TCEs). Specialty films are also required for particular applications such as switchable optical materials, films that have a high transmittance in the photosynthetic spectral region with a high reflectance or absorptance elsewhere, and ultraviolet-reflecting or absorbing interference films. As important elements of the space program for the control of spacecraft temperature, within the passive category, optical films have received the most attention of scientists and thin-film technologists.

Applications for protective coatings are numerous and can be divided into physical (erosion) and chemical (corrosion) protective films. Both erosion and corrosion protection are important concerns for turbine blades, and applications unique to solar energy technologies are determined by a combination of environmental, thermal, and rotational speed conditions. Low-velocity–low-temperature applications include leading-edge protection of wind turbine blades and very large turbine blades for low-temperature heat engines, such as those proposed for open-cycle ocean thermal gradient conversion and ocean current energy conversion devices. High-velocity–high-temperature applications include turbine blade protection for open and closed Brayton cycle heat engines. Corrosion protection is required for low- and high-temperature heat exchanger surfaces and for front- and second-surface mirrors. Photovoltaic devices require encapsulation for both erosion and corrosion protection.

Surfaces with specific polar and dispersive energies may be useful in enhancing the performance of thermal conversion devices. A high-polar-energy heat exchanger surface may improve the wetting of heat-exchange fluids and thus increase the thermal transfer coefficient. A low-surface-energy coating on the reflecting surface of a mirror reduces the adhesion of dust particles in the presence of moisture, promising reduced maintenance costs and extended useful mirror lifetime.

Selective membranes are being employed in advanced solar energy conversion devices. Ionic heat engines and batteries require durable high-temperature membranes. Osmotic head hydroelectric systems are proposed to operate at a brine–fresh water interface and will require membranes that can endure high pressure and be cleaned repeatedly.

This chapter provides a closer look at each of the functions of passive thin films outlined here and describes the materials state of the art in each instance. A brief discussion of the taxonomy of solar energy conversion devices is provided in Section 9.2 as background for the thin-film applications (Sections 9.3–9.6).

9.2   SOLAR ENERGY CONVERSION DEVICES

The exciting field of solar energy research and development has evolved to include the extraction of energy from a wide range of secondary natural energy systems in addition to the direct conversion of incident sunlight into heat, fuels, and electricity. These secondary natural energy sources include ocean thermal and salinity gradients, waves, tides, ocean currents, hydropower, wind, and processed biomass. Systems may be classified by end use as indicated in Table 9.1.

Applications requiring a thermal input include hot water, heating and cooling of buildings (to 200°C), and process heat for agricultural and industrial purposes (to greater than 1000°C). Solar energy conversion systems that can provide this heat include solar ponds; flat-plate collec-

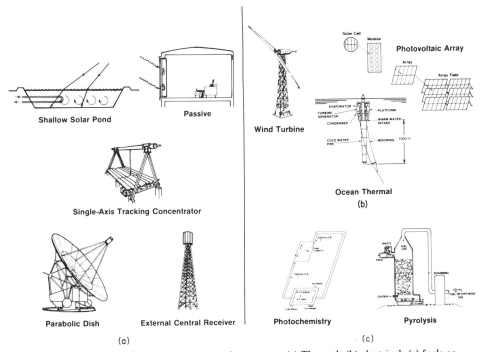

Fig. 9.1   Selected solar energy conversion systems. (a) Thermal; (b) electrical; (c) fuels or chemicals.

TABLE 9.1

Summary of Solar Energy Conversion Systems and Passive Thin-Film Applications

End Use	System	Unit Size	Status	Thin-film applications
Thermal	Passive building design	—	Commercial	Glazing AR coatings, absorbers, heat mirrors, solar shades
	Solar pond	To 1 MW$_{th}$	Commercial	Enhanced transmission covers, evaporation suppression
	Flat plate	~1 kW$_{th}$ (modular)	Commercial	Selective absorber, glazing AR coatings, heat mirrors
	Augmented flat plate, compound parabolic concentrator (CPC), V trough, evacuated tube	To 5 kW$_{th}$ (modular)	Early commercial	Selective absorbers (to 200°C), glazing enhancement, low-concentration reflectors, heat mirrors
	Parabolic trough, cylindrical trough, segmented linear array tracking systems (SLATS)	To 25 kW$_{th}$ (modular)	Early commercial	Transparent receiver covers, medium-concentration-ratio reflectors, selective absorbers
	One- and two-axis tracking Fresnel lens systems	To 25 kW$_{th}$ (modular)	Early commercial	UV protection of plastic refractors, selective absorbers
	Spherical reflector tracking absorber	To 1 MW$_e$	Developmental	
	Linear central receiver	To 50 MW$_e$	Research	Selective absorbers (to 700°C), nonselective absorbers (> 700°C), high-concentration-ratio reflectors
	Parabolic dish	To 50 MW$_e$ (modular)	Early commercial	
	Point focus central receiver	10–400 MW$_e$	Developmental	
	All photothermal convertors	—	—	Dust-mitigating films, working fluid–container interface protection

Electrical–Mechanical			
Wind energy conversion systems	1 kW$_e$ – 2 MW$_e$ (modular)	Commercial to developmental	Turbine blade protection
Photovoltaics	100 W$_e$ (modular)	Commercial	Encapsulants, AR coatings, transparent conducting electrodes
Heat engines (Rankine, Brayton Stirling, sodium ion)	10 kW$_e$ – 100 MW$_e$	Commercial to developmental	Heat exchanger augmentation, working fluid–container interaction (steam, molten metal, molten salt), membranes
Ocean thermal energy conversion (OTEC)	40 MW$_e$ – 400 MW$_e$	Research	Heat exchanger augmentation– antibiofouling, large turbine blade protection
Osmotic head hydroelectric (salinity gradient)	To > 100 MW$_e$	Research	Selective physical membranes
Tidal	To > 100 MW$_e$ (modular)	Commercial	Seawater corrosion, biofouling protection
Wave energy	1 MW$_e$ (modular)	Developmental	
Ocean currents	To > 100 MW$_e$ (modular)	Conceptual	Ion-separating membranes
Photoelectrochemical	—	Research	
Fuels and Chemicals			
Solar thermal heat	Modular	Research	Feedstock process container protection
Processed biomass: extraction, digestion, fermentation, and distillation, gasification, pyrolysis, liquefaction, combustion	Modular (to > 50 MW$_e$)		Corrosion protection for containers in reducing atmospheres with wide range of contaminants, membranes

tors; evacuated tube collectors; nonimaging, low-concentration-ratio collectors; single-axis tracking concentrators (Fresnel, parabolic, and hemispherical); and two-axis tracking concentrators (parabolic dish, spherical dish, and central receiver).

Systems for converting sunlight to electrical (or mechanical) outputs may do so directly (e.g., photovoltaic and photoelectrochemical) or indirectly (e.g., solar thermal input to a heat engine and wind energy conversion). Such devices may be modular and distributed (e.g., parabolic troughs and photovoltaic arrays) or large and centralized (e.g., solar thermal central receiver and ocean thermal gradient closed Rankine cycle heat engine).

Fuels and chemicals from solar energy may be obtained from the processing of biomass residues or energy plantation products, from the synthesis of materials using solar thermal or solar electric energy, and from direct conversion using photochemical or photobiological systems.

Figure 9.1 illustrates some diverse solar energy conversion systems.

9.3   OPTICAL FILMS

Optical films have received considerable materials research and development support during the past 25 years both within and outside the national solar energy program. Indeed, only the study of photovoltaics has had greater emphasis by solid-state physicists and material scientists than the development of selectively absorbing films with high solar absorptance and low thermal emittance. In addition to solar absorbers, other optical films are required in photothermal converters. Reflector materials are needed in concentrating collectors. Antireflecting coatings improve the performance of glazings, encapsulants, and photovoltaic devices. Heat mirrors can control the temperature of building spaces and greenhouses, improve heat trapping in solar collectors, and improve furnace performance where thermal processing of materials is required. Transparent conducting materials are essential as front electrodes on Schottky-barrier and metal-insulator semiconductor (MIS) photovoltaic devices. Ultraviolet-reflecting or absorbing interference filters protect plastic glazings and encapsulants from photodegradation. Switchable films are desirable for the protection of plastic solar collectors from high temperatures in the case of stagnation (loss of coolant). Films that transmit in the photosynthetic spectral region and absorb or reflect at all other wavelengths have been suggested for more effective management of greenhouse heating and cooling loads.

The complete optical characterization of thin films requires measurement of the spectral bidirectional reflectance and transmittance from 0.35

to 30 μm over all angles of incidence and exitance. Such a thorough characterization, however, is unnecessary in all but the most specialized of applications. In discussing low-temperature applications of optical films, the electromagnetic spectrum can be conveniently split into the solar region (0.35–2.5 μm) and the thermal infrared region (2.5–30 μm). Four independent parameters suffice to define optical performance: the solar reflectance ρ_s and transmittance τ_s and the thermal reflectance ρ_t and transmittance τ_t. These four parameters are determined by weighting the spectral directional hemispherical transmittance and reflectance with the solar (E') and blackbody (I_{bb}) distribution functions over wavelength as follows:

$$X_s \equiv \int X(\lambda; \theta, \phi; 2\pi) E'_\lambda \, d\lambda \Big/ \int E'_\lambda \, d\lambda \tag{9.1}$$

$$X_t \equiv \int X(\lambda; \theta, \phi; 2\pi) I_{bb\lambda}(T) \, d\lambda \Big/ \int I_{bb\lambda}(T) \, d\lambda \tag{9.2}$$

(for $X \equiv \rho$ and τ) where E'_λ is the solar irradiance from Thekaekara [1] corrected to air mass 1.5, $I_{bb\lambda}(T)$ is given by the Planck distribution $C_1\lambda^{-5}(e^{C_2/\lambda T} - 1)$, where C_1 and C_2 are constants equal to 3.74×10^{-16} W/m^2 and 1.44×10^{-2} m K, respectively, and θ and ϕ are the angles specifying the incident radiation direction with respect to the optical surface assumed to be near normal unless otherwise specified. The solar absorptance α_s and the normal thermal emittance ϵ_t can then be determined from the equalities

$$\alpha_s + \rho_s + \tau_s = 1 \tag{9.3}$$

and

$$\epsilon_t + \rho_t + \tau_t = 1 \tag{9.4}$$

Thorough discussions of the optical properties and measurements appropriate for solar applications are presented in Touloukian [2] and Masterson *et al.* [3].

As is apparent from Fig. 9.2 the solar and thermal quantities are independent only at low temperatures. At elevated temperatures the spectral overlap between the solar and blackbody distribution functions necessitates the introduction of a fifth parameter λ_c, the wavelength at which an ideal step function change in an optical parameter should occur to optimize simultaneously the solar and thermal properties. This parameter is a function of the system operating temperature and concentration ratio [4] and is approximately 2 μm at room temperature for a concentration ratio of 1.

Ideal solar optical materials are achieved when the properties $\alpha(\epsilon)$, ρ, and τ approach the extremes of 1 and 0 independently in the solar and

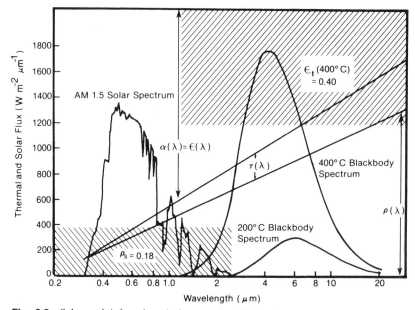

Fig. 9.2 Solar and infrared optical properties. The solar (s) and infrared (t) optical properties of a material are determined by weighting the spectral hemispherical reflectance $\rho(\lambda)$, transmittance $\tau(\lambda)$, and absorptance $\alpha(\lambda)$ (or emittance ϵ) over the solar and infrared radiation distribution functions E'_λ and $I_{bb\lambda}$, respectively. The smoothly varying optical properties of a hypothetical material are illustrated with $\alpha(\lambda) + \tau(\lambda) + \rho(\lambda) = 1$. In addition to the ρ_s and ϵ_t values shown solar and blackbody weighting yields $\alpha_s = 0.77$, $\tau_s = 0.05$, τ_t (400°C) = 0.13, and ρ_t (400°C) = 0.47.

infrared spectral regions. Table 9.2 lists the ideal values for the applications described previously. Although the overlap of the blackbody and solar spectra at elevated temperatures introduces some compromises, this tradeoff alone is not severe. As an example, based on this overlap a solar absorptance of 0.95 or greater requires that the thermal emittance be no less than 0.12 at 1000°C. However, the additional effect of thermal broadening of the transition from low to high reflectance at elevated temperatures and the constraints of the Kramer–Kronig relationships on the real and imaginary components of the complex index of refraction (n and k) introduce a more modest fundamental limit for ϵ_t. Including these effects, Trotter and Sievers [5] have calculated that for $\alpha_s \geqslant 0.96$ ϵ_t must be no less than 0.25 at 600°C.

Although considerable work remains to improve the optical properties, cost, and durability of films for particular applications, much of this development will be evolutionary, emphasizing and refining present materials and techniques. A research direction of potential long-range impor-

TABLE 9.2

Ideal Optical Properties for Materials in Solar Applications

Optical Film	Ideal Solar (s) and Infrared (t) Properties					
	α_s	τ_s	ρ_s	ϵ_t	τ_t	ρ_t
Selective absorber	1.0	0.0	0.0	0.0	0.0	1.0
Reflector	0.0	0.0	1.0	0.0	0.0	1.0
Heat mirror	0.0	1.0	0.0	0.0	0.0	1.0
Antireflection coating	0.0	1.0	0.0	0.0	1.0	0.0
Transparent electrode	0.0	1.0	0.0	0.0	0.0	1.0
Switchable film						
Cool season (heating)	0.0	1.0	0.0	0.0	0.0	1.0
Hot season (cooling)	0.0	0.0	1.0	0.0	0.0	1.0
Solar shade	0.0	0.3–0.7	0.7–0.3	0.0	0.0	1.0
Greenhouse glazing	0.0	0.5^a	0.5	0.0	0.0	1.0

^{a} 1.0 for $\lambda < 0.75$ μm.

tance is the development of switchable materials (thermally, electrically, or photon activated). A window coating, for example, might be switched from a heat mirror state ($\tau_s \approx 1$) for optimal winter use to a solar shade ($\tau_s \approx 0.3$, $\rho_s \approx 0.7$) to limit summertime cooling loads. Such materials represent an exciting frontier for materials engineering; however, it is important to note that relatively unexciting criteria such as cost, maintenance, and durability will ultimately determine the marketability of such area-intensive films. Solar devices thrive on simplicity, and well-established if cumbersome alternatives exist in many instances which add to the established market inertia for new materials. For window coatings the competition takes the form of shades, curtains, and awnings.

1. Absorber Surfaces

The optical properties—solar absorptance α_s and thermal emittance ϵ_t —of the receiver surface are important in a wide range of photothermal conversion devices, from passive solar buildings to sophisticated two-axis tracking concentrators. Primary references to absorber surface research and development number in the thousands, and the field has been reviewed extensively [6–10].

A partial list of materials in various stages of research, development, and commercialization is included in Table 9.3. The selective materials exploit the physical effects listed in Table 9.4, either singly or in combination, to achieve simultaneously a high solar absorptance and a low thermal emittance. Of the commercial materials, electrodeposited black chrome ($\alpha_s \approx 0.95$, $\epsilon_t \approx 0.10$) dominates the selective absorber market to 200°C, and improved control of the Cr^{3+} concentration in the plating bath

TABLE 9.3

Properties of Selected Solar Absorber Surfaces[a,b]

Material	Technique	Supplier(S)/Developer(D)	Maturity[c]	α_s	$\epsilon_t(T)$	T Stability[d] (°C)	Issues	Reference
Black chrome	Electrodeposited	Many	5 (low T) / 4 (high T)	0.94–0.96	0.05–0.10 (100) / 0.20–0.25 (300)	300		[6]
Pyromark	Paint	Tempil	5	0.95	0.85 (500)	<750	Adhesion, curing process	[6]
S-31 (nonselective)	Paint	RockwellInternational International	5	0.8–0.85	0.8–0.85	>550		[6]
Solartex	Electrodeposited	Dornier (W. Germany)	5	0.93–0.96	0.14–0.18 (310)	700		Product literature
Solarox (proprietary)	Electrodeposited	Dornier (W. Germany)	5	0.92	0.20	200		Product literature
Black epoxy	Paint	Amicon Corp.	5	NA	NA	NA		Product literature
436-3-8	Paint	Bostik (U.S.M. Corp.)	5	0.90	0.92	NA		Product literature
Enersorb	Paint	Desoto	5	0.96	0.92	NA		Product literature
7729	Paint	C. H. Hare	5	0.96	0.90–0.92	NA		Product literature
R-412	Paint	Rusto-leum Co.	5	0.95	0.87	NA		Product literature
5779	Paint	Rusto-leum Co.	5	0.95	0.90	NA		Product literature
Nextel (nonselective)	Paint	3M	5	0.97–0.98	>0.90	150		Product bulletin 206
Novamet 150 (proprietary)	Paint	Ergenics	5	0.96	0.84	800 (1 h)		Product literature
Maxorb	Proprietary	Ergenics	5	0.97 (±.01)	0.10 (±.03)	150 (20 weeks), <400 (1 h)		Product literature
Tabor black (NiS–ZnS)	Electrodeposited and overcoat	Miromit	5	0.91	0.14			Product literature

Material	Method	Manufacturer	Maturity[c]	Absorptance	Emittance (temp)	Temperature stability[d]	Notes	Reference
Proprietary Al$_2$O$_3$–Mo–Al$_2$O$_3$ (AMA)	Vacuum-deposited Sputtering evaporation	GE Honeywell	5 4	0.85–0.95	0.34 (100), 0.11–0.4	> 550		[11] [10]
Multilayer (proprietary)	Proprietary	OCLI	4	0.95	0.05 (100)	> 300		[6]
NiS–ZnS	Two-Layer electroplated	Many	3	0.96	0.07	< 250	Moisture-sensitive	[10]
Proprietary inorganic	Paint	Martin Marietta	3	0.9–0.95	0.9–0.95	> 550	Adhesion curing	[6]
Selective paint	Dip-coated	Honeywell	3	0.92	0.13 (100)	> 150		[12]
Si/Ag bilayer with AR	CYD	U. of Arizona	3	0.80	0.07 (500), 0.05 (100)	500 (vacuum)		[12]
CuO–Ag–Rh$_2$O$_3$	Fired organo-metallic paint	Englehard	3	0.9	0.1	< 400		[12]
CuO–Au	Fired organo-metallic paint	Englehard	3	0.8	0.06	> 600		[12]
Silicone and silicate	Paint	EXXON	3	0.98	0.9	> 700		[12]
Au–Al$_2$O$_3$ (cermet)	Sputtering	U. of Sydney	2	0.95	0.025 (20)	< 300		[13]
Cu–Al$_2$O$_3$ (cermet)	Sputtering	U. of Sydney	2	0.91	0.045 (20)	< 200		[13]
SnO$_2$:F–black enamel	Spray	University of Delft (the Netherlands)	2	0.92	0.15	200	Very rugged[14]	

[a] From Call [6].
[b] NA, not available; AR, antireflecting.
[c] Maturity: development status of absorber surfaces—5, commercial; 3, development; 1, research.
[d] Temperature stability for most absorber surfaces is poorly defined and depends critically on exposure environment, conditions of test, and survival criteria.

(cont.)

TABLE 9.3 (*cont.*)

Material	Technique	Supplier(S) Developer (D)	Maturity[c]	α_s	$\epsilon_t(T)$	T Stability[a] (°C)	Issues	Reference
Tungsten–stainless steel dendrites	Sputtering chemical conversion	IBM	2	0.95		> 550	Angle of incidence	[10]
Gold smokes	Gas evaporation	Many	1	0.99	0.1 (low T)	100	Fragile	[10]
Germanium	Gas evaporation	Many	1	0.91	0.2 (160) 0.5 (250) 0.8 (200)			[10]
	Paint silicone binder	Many	1	0.91				
Au–MgO cermet	Etched Rf sputtering	Penn State U. MIT Lincoln Labs	1 1	0.93	0.1	< 400 < 300		[15] [16]
Cr–Cr$_2$O$_3$ cermet	Rf sputtering	MIT Lincoln Labs	1	0.92	0.08	< 400		[16]
SiO–Cr–SiO multilayer	Vacuum evaporation		1	0.88	0.1 (low T)	< 450		[10]
Ni–Al$_2$O$_3$ cermet	Evaporation	Cornell U.	1	0.94	0.16 (100), 0.35 (500)	500		[12]
Pt–Al$_2$O$_3$ cermet	Vacuum evaporation	Cornell U.	1	0.94	< 0.3 (500)	600		[12]
α-Silicon	Plasma discharge	AML–BML	1	> 0.9 (calc.)		> 500		[12]
Si–CaF$_2$ and Ge–CaF$_2$ cermets	Sputtering	RCA	1	0.7	< 0.1			[12]

Material	Process		α	ε (T)	T (°C)	Comments	Ref
Al₂O₃–ZrCₓNᵧ–Ag	Reactive sputtering	1	0.91	<0.05	175 (air), 700 (vacuum)		[12]
Aluminum	Anodized (organic dye) — Albany, or Bureau of Mines	1	0.96	0.98 (350)	<350		[10]
	Anodized (KMnO₄ dye)	1	0.80	0.35 (<100)			[10]
PbS	Vacuum-deposited	1	0.98	0.05 (100), 0.2 (240), 0.3 (300)	300	UV and O₂ stability	[10]
	Paint (silicone binder)	1	0.94	<0.8 (200)	Binder thickness		
Cu₂S	Chemical conversion	1	0.79	0.2 (200)			[10]
WC + Co	Plasma spray	1	0.95	0.28 (200), 0.4 (600)	>800		[10]
Cr₂O₃ + Co	Plasma spray	1	0.9 (800)	0.5 (800)	>800		[10]
Co₃O₄	Electroplated	1	0.9	0.3 (140)	>1000		[10]
304 stainless steel	Chemical conversion	1	0.91				[10]
	Thermal oxide (760°C)	1	0.82	0.15 (100), 0.2 (300)			[10]
Inconel and incoloy	Thermal oxide (1000°C)	1	0.85–0.90		1000		[12]
Steel (Fe₃O₄)	Chemical conversion	1	0.90	0.07 (90), 0.35 (200)			[10]

TABLE 9.4

Generic Types of Selective Absorber Surfaces[a]

Single material with ideal intrinsic solar absorptance/IR
  reflectance [HfC, ReO_3]
Bilayer (infrared reflector–solar absorber tandem) [Si/AG]
Multilayer (interference) [Al_2O_3–Mo–Al_2O_3]
Surface topography (light-trapping morphology with physical dimension
  approximately equal to the wavelength of the solar spectrum) [tungsten dendrite]
Small-particle effects (Mie Scattering, resonance, and dielectric
  anomalies) [Pt–Al_2O_3 cermet]

[a] Material examples appear in brackets. (From Call [6].)

TABLE 9.5

Issues of Concern for Absorber Surfaces[a]

Operating efficiency
  High $\alpha_s(T)$
  Low $\epsilon_t(T)$
  Angle of incidence effects
  High thermal conductivity
Operating life and degradation mechanisms
  Temperature stability [maximum operating temperature, gradient,
    transients (shock), and cycling (fatigue)]
  Effects of solar photon flux (ultraviolet)
  Impact and abrasion resistance (dust and hail)
  Effect of adherent dust on optical properties
  Chemical stability (atmospheric, working fluid, and system
    contamination, rain, humidity)
  Vacuum stability
Repairability
Cost per unit area
Materials resource limitation or vulnerability
Geometric constraints in application of coating
Shaping or forming after coating
Limitations on substrate candidates

[a] From Call [6].

indicates that thermal stability to 400°C may be achieved [17]. Electro-deposited black nickel with a protective overcoat for impeding moisture penetration and electrodeposited black cobalt for applications to 400°C are also being produced. A number of nonselective ($\alpha_s \approx 1$, $\epsilon_t \approx 0.9$) paints are commercially available, including Pyromark, which has undergone extensive tests as the baseline absorber surface for the Barstow, California, 10 MW_e solar thermal pilot plant. Issues of concern to absorber surface users are not limited to optical performance and thermal stability, as is emphasized in Table 9.5.

TABLE 9.6

Importance of Reducing Receiver Thermal Emittance to Solar Thermal Systems[a]

System	T_1 (°C)	A_c/A_r	$\beta^b \equiv \dfrac{\sigma(T_1^4 - T_2^4)}{\phi\tau E\rho A_c/A_r}$
Passive	25–80	1	0–0.7[b]
Flat plat	40–180	1	0.13–3.7[c]
Evacuated tube (CPC, V troughs)	100–200	1.5–3	0.5 –3.3
Linear Fresnel and cylindrical trough	150–250	6–10	0.3 –1.3
Linear parabolic and segmented linear array tracking system (SLATS)	200–350	20–70	0.07–0.77
Two-axis tracking Fresnel	200–350	50–100	0.05–0.3
Spherical reflector tracking absorber	300–500	100–200	0.06–0.4
Two-axis tracking parabolic dish	500–1100	1000	0.04–3.5
Central receivers			
Steam cycle (external receiver), Barstow 10 MW$_e$	500	400	0.09
Steam cycle (external receiver), 100 MW$_e$	500	1200	0.03
Advanced molten salt and metal (external receiver)	600	1800–2500	0.03–0.04
Advanced open and closed advanced Brayton cycles (cavity receiver)	1000–1400	2500–4000	0.06–0.3

[a] Assumes $\tau = 0.83$, $E = 1.0$, $\rho = 0.85$, $\phi = 790$ W/m², $T_2 = 27$°C.

[b] The parameter β is defined as the radiated heat flux of a blackbody at the system operating temperature T_1 divided by the incoming solar flux delivered by the collector to the receiver surface. The parameters are the solar flux ϕ, the solar transmittance of the system optics τ, the spillover factor E, the solar reflectance of the concentrator ρ, and the geometric concentration ratio A_c/A_r. For $\beta > 1$ the operating temperature cannot be obtained unless the receiver thermal emittance ϵ_t is less than the receiver absorptance α_s divided by β. The two classes of receivers are external and cavity geometries. In the external geometry the solar radiation is received on the external surface of a body (hemisphere, sphere, tube, etc.). In a cavity geometry the solar radiation is absorbed on the interior surface of the body, incident through an aperture in the body. (From Call [6].)

[c] $\rho = 1.0$.

For purposes of comparison, the receiver thermal efficiency η of a system can be described by a simplified expression

$$\eta = \alpha_s - \beta\epsilon_t \qquad (9.5)$$

where β, a measure of the importance of the receiver thermal emittance to the application, is the radiated heat flux of a blackbody at the system operating temperature T divided by the incoming solar flux delivered by the collector to the receiver surface. Table 9.6 lists β for representative applications as a function of operating temperature and system concentration factor A_c/A_r. It is somewhat paradoxical that, in general, β is highest for current designs of low-temperature systems because the operating temperature of high-concentration-ratio systems is limited by other material constraints, such as the interaction of the working fluid and the container walls.

2. Reflecting Films

To exploit the 5800 K thermodynamic potential of the sun using photothermal systems it is necessary to concentrate the sunlight. The degree of concentration is limited ultimately by the finite dimensions of the sun, which impart a 9-mrad divergence to unscattered incoming rays. In practical cases for solar energy conversion, this limit is not approached because of the optical quality of affordable mirrors and the inability of receiver and containment materials to survive temperatures above approximately 2000 K.

Although the hemispherical reflectance is an important measure of mirror acceptability, as the concentration ratio is increased, the specular reflectance (reflectance within a solid angle of exitance equal to a narrow solid angle of incidence at the specular angle) is the appropriate figure of merit. Whereas the hemispherical reflectance is determined primarily by the electronic and topographical features of the reflecting surface and its protecting layers, the specular reflectance is also a function of the microstructural and configurational aberrations of the mirror (both short- and long-wavelength variations from the ideal). These aberrations are properties of the entire reflector structure including the substrate or superstrate, bonding layers, protective layers, and structural support elements as illustrated in Fig. 9.3 for the two geometries of reflector: superstrate and substrate.* (The substrate mirrors include mirrors with thin protective oxides or coatings on the surface of the reflector.)

For reasons that are clear from Fig. 9.4, only two metals, silver and aluminum, are commonly considered for the reflecting layer itself. Freshly deposited bare silver has a high solar reflectance ($\rho_s \approx 0.98$) but does not form a protective oxide layer and thus has poor resistance to environmental attack. Silver that has been protected with a thin overlayer (of refractive index $n \sim 1.5$) has $\rho_s \approx 0.97$. The reflectance of aluminum is reduced through most of the visible spectral region (with a pronounced dip at 0.8 μm) because of interband electronic transitions in the crystalline solid. However, the reflectance at wavelengths shorter than 0.35 μm is greater than that of silver, and a free-electron (Drude model) calculation for aluminum which eliminates interband transitions yields a solar reflectance of 0.99 as compared to a measured value of 0.92 (0.88 with a thin

---

*Historically, the two geometries have had the somewhat misleading designation of second-surface and front-surface mirrors, respectively.

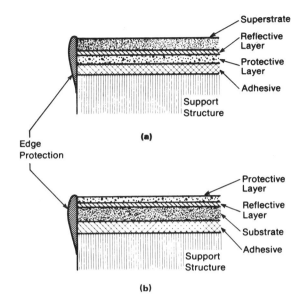

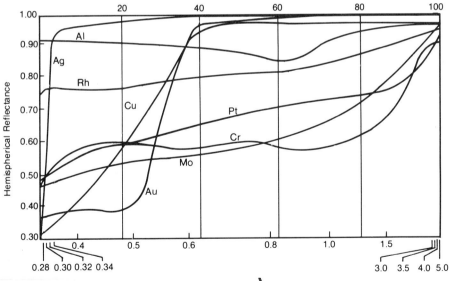

Fig. 9.3 Schematic diagram of (a) superstrate and (b) substrate reflectors. (From Masterson [3].)

Fig. 9.4 Hemispherical reflectance of selected candidate mirror metals. (Adapted from Touloukian [2].)

TABLE 9.7

Material	Product example and manufacturer	Index of refraction	τ_s	τ_r	Typical thickness (nm)[b]	Density (g cm^{-3})
Acrylics	Plexiglas (Rohm and Haas)	1.49	0.90	0.02	3.2	1.20
Halocarbons	FEP Teflon (Dupont)	1.34	0.96	0.26	0.05	2.1
	PVF Tedlar (Dupont)	1.46	0.92	0.21	0.10	1.4
Silicones	RTVs (GE)					
Polyesters	Mylar (Dupont)	1.64	0.87	0.18	0.13	1.4
Polycarbonates	Lexan (GE)	1.59	0.84	0.02	3.2	1.2
Borosilicate glass	7806 fusion (Corning)	1.47	0.88	0.02	1.1	2.1–2.5
Soda lime glass	Float (Ford)	1.51	0.85	0.02	3.2	2.5

[a] Data compiled from Hampton and Lind [22], Ratzel and Bannerot [23], Bouquet [24], Modern Plastics Encyclopedia [25], and Van Krevelen [26].
[b] Value used to obtain optical properties.

protective layer) [18], prompting discussion that a room temperature-stabilized amorphous phase of aluminum might be superior to silver in cost, durability *and* performance [19]. Another approach to an improved reflecting material is the alloying of silver and aluminum by coevaporation or cosputtering.* Some system designs employ a secondary concentration scheme where the secondary reflector may experience an incident flux in excess of 1 MW/m^2 and achieve a temperature of 500°C or higher. High-temperature alloys of silver[†] and thin aluminum films on stainless steel have been proposed for these applications.

High-quality commercial silver mirrors are produced by a wet chemistry electrodeless process and also by evaporation onto glass. Aluminum and silver are both vacuum deposited onto plastic films, and a polished aluminum sheet with an anodized protective film is also available commercially; however, the specularity of such mirrors is a concern. An overview

*Work being performed under contract to the Solar Energy Research Institute by Rockwell International, Rocky Flats Plant.
†Work being performed under contract to DOE Solar Thermal Test Facility Users Association by Richard Zito, Physics Department, University of Arizona.

Properties of Selected Transmitting Materials for Encapsulants and Glazings[a]

Specific heat $(J/°C_1 g)$	Coefficient of thermal expansion $(°C \times 10^{-5})$	Service temperature (°C)	UV resistance[c]	Remarks
0.75	6.10	120	Excellent	Thermoplastic and thermoset; flammable; brittle with age or cold; long-term field tests show good weatherability of optical properties
1.17	10–16	200	Excellent	high cost;
1.26	5	100	Good–excellent	nonwettable surface can cause cleaning problems
	0.5–1 (laminates)	150–200	Good	High water permeability; soft surface; not very specular
1.05	1–4	100–150	Poor	May become brittle without UV protective coating
1.19	7	120–130	Fair–good	Specular transmittance; high impact strength; brittle after outdoor aging
0.75	0.9	820 (softening)		
	0.3	700 (softening)	Excellent	

[c] The UV resistance is a subjective evaluation of material performance based on field weathering experience as well as UV laboratory testing. Additives in materials as well as surface ptotective UV absorbers may drastically alter material performance.

of current solar reflector research and development can be found in Lind and Ault [20], and a thorough review of mirror technology is found in Schweig [21].

The superstrate or substrate to which the reflecting layer is attached is a major factor in the performance and durability of the mirror and may be a glass or a polymeric material. Hampton and Lind [22] review these materials in some detail, and a summary of the appropriate properties is included in Table 9.7. Many of these same materials are being used as transparent glazings for flat-plate collectors, solar ponds, and greenhouses, and as encapsulants for photovoltaic modules. Thus, some of the additional properties appropriate to these applications are included.

3. Antireflecting Coatings

The surface reflection of light as it encounters a change in index of refraction at the boundary between two materials can be a critical loss in solar applications such as glazings, photovoltaics, and absorber surfaces.

At normal incidence the surface reflectance is given by

$$R = \left[(n_1 - n_2)^2 + (k_1 - k_2)^2 \right] / \left[(n_1 + n_2)^2 + (k_1 + k_2)^2 \right] \qquad (9.6)$$

where n and k are the real and imaginary components of the complex index of refraction $\tilde{n} \equiv n + ik$ for materials 1 and 2 on either side of the interface. R is approximately 0.04 for a glass–air interface. For high index of refraction materials such as some semiconductors (e.g., silicon), losses in excess of 0.35 can be expected if no attempt is made to reduce the surface reflectance.

Coatings or treatments for reducing the surface reflectance loss are referred to as antireflecting layers. They are formed by interposing a layer of intermediate index of refraction between the high- and low-index materials. The simplest technique consists of depositing an antireflecting layer of index of refraction n and thickness d determined by

$$d = \lambda/4n; \qquad n = \sqrt{n_1 n_2} \qquad (9.7)$$

where λ is the wavelength of minimum desired surface reflectance. A value of 550 nm, near the peak of the terrestrial solar spectrum, is typically chosen for λ as a value that will minimize solar losses at glazing and nonselective absorber interfaces. However, a more careful analysis of optimum λ may be required for photovoltaics and selective absorbers where the photoresponse is likely to be a more complicated function of wavelength.

A variety of inorganic dielectrics deposited by sputtering, evaporation, glow discharge, chemical vapor deposition (CVD), and thermal oxidation have been explored as potential antireflecting coatings. In particular, Si_3N_4 has received considerable attention as a coating for silicon [27]. Etching the surface of a material reduces the index of refraction in the etched region; and the effect of the etching process can be modeled through an effective medium approach [28]. The surface reflectance of glass has been reduced in this manner, with the major tradeoff being increased dust adhesion due to the larger surface area, leading to greater cleaning problems.

Researchers [27, 29] have explored theoretically and experimentally the notion of a graded-index, antireflective coating. An experimental graded silicon oxynitride film is illustrated in Fig. 9.5. The importance of such graded films is to broaden the wavelength interval over which the reflectance is reduced and to expand the effective aperture for the antireflective coating (i.e., extend the reduction in reflectance to angles of incidence farther from normal).

An abbreviated list of inorganic dielectrics with thin-film values for the real part of the index of refraction n is included in Table 9.8.

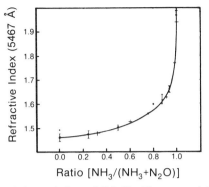

Fig. 9.5 Refractive index variations of SiO_xN_y. The oxygen/nitrogen ratio in a SiO_xN_y film can be continuously varied by changing the ratio of inflowing gases (NH_3–N_2O) in a chemical vapor deposition process with helium, nitrogen, and SiH_4 constituents at 700°–885°C. Thus, the index of refraction n can be varied between the expected values for SiO_2 and Si_3N_4. (From Seraphin [27].)

TABLE 9.8

Index of Refraction n of Selected Thin Inorganic Films [a]

Material	n [b]
Ta_2O_5	2.8 (0.4)
TiO_2	2.4–2.9 (0.4)
CeO_2	2.3 (0.4)
WO_3	2.29 (0.55)
ZrO_2	2.10 (0.4)
PbF_2	1.75 (0.4)
Al_2O_3	1.62 (0.4)
SiO_x	1.41–1.48 (0.4)

[a] Values will change depending on deposition conditions, approaching bulk values for the same material as the density is increased. (Data from Hussman and Kerner [30] and DiStefano *et al.* [31].)

[b] Wavelength (μm) is given in parentheses.

4. Heat Mirrors

Heat mirrors are films that possess a high infrared reflectance (and thus a low infrared emittance). For solar applications a second constraint is placed on a candidate material: high solar transmittance. Heat mirrors may be incorporated into greenhouse glazings and architectural windows to enhance the thermal gain of the structures. Applications to flat-plate, evacuated tube, or solar pond heat collectors allow good optical selectivity without placing stringent thermal stability requirements on the material (in contrast to a system that gains optical selectivity at the absorber surface). This relaxation of the material thermal stability requirement is likely to be even more significant for very high-temperature applications such as Brayton cycle systems, where the receiver wall may be 1400°C, but a heat mirror located at the aperture of the cavity receiver could possibly be cooled to 800–900°C. Alternatively, heat mirrors may be applied directly to an absorbing surface to form an inverted tandem stack such as tin oxide doped with fluorine on black enamel [14]. An indirect application of heat mirrors to solar energy technologies occurs in the processing of materials (e.g., semiconductors for photovoltaic devices) in high-temperature furnaces where a heat mirror may greatly enhance the energy efficiency of the processing step without compromising visual access [32].

Heat mirror materials of interest in solar applications are of three principal types: thin metallic films, conducting metallic oxides, and multilayer stacks. Thin metallic films of interest have been Cu, Ag, and Au, all of which have high IR reflectance and a modest transmittance in the solar spectrum. Heavily doped n-type SnO_2, In_2O_3 (including indium tin oxide) and the metallic oxides ReO_3, Na_xWO_3, and Cd_2SnO_4 have been investigated as heat mirrors [33]. Multilayer films that have been studied include TiO_2–Ag–TiO_2 [34] and ZnS–Ag–ZnS [35]. The dielectric layers in the multilayer systems act as antireflection coatings for the solar spectrum. The inner dielectric layer also plays a critical role in determining the microstructure of the metallic layer. A summary of the optical properties of heat mirrors is included in Table 9.9.

TABLE 9.9

Solar Transmittance τ_s and Infrared Reflectance ρ_t
of Selected Heat Mirrors and Solar Shades[a]

Material	τ_s	ρ_t
ZnS–Ag–ZnS	0.65	0.94
Cd_2SnO_4	0.89–0.78	0.8–0.9
TiO_2–Ag–TiO_2	0.54	0.98
Sn-doped In_2O_3	0.90	0.92
In_2O_3	0.84	0.86
Cu–Ag (electrodeless)	0.13	0.95 (9.7 μm)
Co_3O_4	0.26–0.67	Not available

[a]Data from Fan and Bachner [34], Koltun and Faizier [35], Haacke [36], Greenberg [37], Rekant and Borisova [38], and Apfel [39].)

5. Transparent Conducting Electrodes (TCEs)

Since the electronic properties of a material determining the IR reflectance also are involved in defining the electric conductivity, the set of materials previously considered as heat mirrors also is of interest as TCEs. In addition to the optical properties, the thin-film electrical sheet resistance and the interface resistance between the underlying electrode and the semiconductor substrate are parameters of interest for TCEs. The boundary resistance is particularly sensitive to the defect density at the interface, and thus the deposition conditions and material compatibility are critical. Vossen [40] has extensively reviewed the material state of the art, concluding that extreme variations in electrical properties (up to 10 orders of magnitude for dc conductivity) can be reported for nominally similar materials, and careful study of the film microstructure and impurity concentrations is needed to define an electrode material. As an example for a specific material, Fan *et al.* [41] report the importance of controlling the oxygen pressure during vapor deposition on tin-doped In_2O_3.

Analogous to the situation for absorber materials (where reductions in solar absorptance represent an irretrievable loss to the system but increased thermal emittance can be at least partially compensated for in system design), the solar transmittance of TCEs is the fundamental parameter of concern, and the electrical properties of a film may be at least partially compensated for by device design [42]. An example is the tradeoff between large-area single cells and multiple parallel Schottky-barrier or metal-insulator semiconductor (MIS) photovoltaic cells. Large-area cells have simple processing and fewer interconnections but require very low and uniform electric resistance, whereas multiple parallel cells can greatly reduce series resistance losses.

A platinum TCE on amorphous silicon (discussed in Chapter 8), prepared from the glow discharge of SiH_4, has yielded a 5.5% efficient photovoltaic device over limited areas [43]. Metal electrodes on polycrystalline silicon are being investigated with a goal of photovoltaic cells with 12% efficiency [44]. Other applications for TCEs may be found in large-area switchable devices, for example, as electrodes for electrochromic or liquid crystal materials in switchable solar shades.

6. Specialized Optical Films

A large number of special applications exists for particular thin optical films in solar technologies. In contrast to most applications described in previous sections, suitable or proven materials do not exist and exploratory research will be required to evaluate the viability of each concept.

For thermal control of buildings and for thermal protection of low-temperature solar collectors, materials whose optical properties can be switched are desirable. To reduce the materials intensity of low-temperature solar collectors, extruded high-strength plastics are being considered. When designed to operate efficiently, such collectors may experience temperatures in excess of 200°C when coolant flow is interrupted under direct sunlight. A straightforward solution to this problem has been proposed in the form of thermally reversible bleaching of organic absorber dyes [45]. The main drawback of this technique is that the solar absorptance and selectivity of such dyes at normal operating temperatures may not allow high efficiencies.

Switchable films at the glazing surface should be considered. Two such techniques might employ electrochromic or liquid crystal technology. Both technologies have been studied extensively from the perspective of visual display devices. For this solar application the electrical requirements for direct switching of large areas may be the limiting factor, and power consumption must be reduced to less than 0.1 mW/cm^2 in the switched state. Field-effect liquid crystal devices that operate between crossed polarizers and require fractions of 1 μW/cm^2 should be explored, and the discovery of a thermally reentrant nematic liquid crystal phase could provide a truly elegant "passive" solution to this problem. Although development of acceptable optical and electrical properties is a monumental task, any material produced will be required to exist for significant lengths of time (> 5 years) in a harsh environment of heat, sunlight, humidity, and pollutants not typical of the hermetically sealed visual display environment.

The protection of plastics used in glazings, bubble domes, and encapsulants from photodegradation by UV-absorbing coatings is an important concern. Interference filters of $ZnS-ThF_4$ and $ZnS-Al_2O_3$ have been studied and found promising for the harsh UV space environment [30]. Additives to commercial plastic materials that scavenge UV reaction products in the presence of air and moisture are well established. The cost effectiveness of coatings for terrestrial use must be judged with respect to both system design modifications that allow easy replacement of plastic components and the performance of available additives.

The photosynthetically active region of the solar spectrum lies at wavelengths below 0.75 μm and comprises only 50% of the air mass 2 insolation. Separation of the photosynthetic and heating portions of the solar spectrum to reduce greenhouse cooling loads and to manipulate independently the delivery of heat to better meet greenhouse heating demands has been proposed using dichroic mirrors in a baffled glazing geometry [46].

9.4 PROTECTIVE COATINGS

Natural and artificial protective layers are essential in many applications to reduce the effects of erosion and corrosion. Depending on the protection needed, the physical (modulus, deformation energy, microstructure, etc.) and/or chemical (adhesion, ionic permeability, etc.) properties of the protective film are important. A surface that is moisture sensitive, for instance, might be protected either by a film that is impermeable to water molecules (e.g., densified SiO_2) or by a film whose interfacial chemistry with the surface excludes water from reacting with that surface (e.g., silicones).

Many reviews of thin-film applications for solar energy ignore the category of protective or passivating layers. This neglect is paradoxical because, although the innovative frontier in materials development may appear to be improved performance, in reality some of the biggest questions facing solar technologies are the durability, reliability, and lifetime of materials and components. In deference to the magnitude of the overall subject of protective coatings, this section will attempt to illustrate some critical protective needs unique to solar energy conversion systems.

1. Protection for Turbine Blades

Turbine blades for solar applications span a wide range of rotational velocities, operating temperatures, and environmental conditions. At low velocities (typically 10–100 rpm) very large-blade structures (up to 100 m in diameter) fabricated from fiberglass or concrete composites are contemplated for advanced wind turbines and for open-cycle ocean thermal energy conversion systems. Edge protection from erosion may be required in both wind and low-temperature heat engine environments. Conventional experience with helicopter rotors, propeller blades, and wind turbine blades has turned to thick-film technology (e.g., metal alloys and rubber) for incident particle shock absorption. For high-efficiency turbine design in solar applications where blade drag and weight may be critical, new applications of dense, adherent, and durable thick-film coatings may appear.

Solar applications for high-temperature turbine blades differ very little from conventional gas turbine applications. However, some system designs that propose purely solar inputs to the turbine would greatly reduce the "hot corrosion" problem experienced at high temperatures in the presence of fuel impurities, especially sulfur. A reduction in hot corrosion for these designs would include a tradeoff with regard to unavoidable enhancement of the thermal or mechanical cycling fatigue as the temperature and/or gas flow fluctuates with the solar input.

The materials state of the art for gas turbine coatings was reviewed in a session of the 1978 International Conference on Metallurgical Coatings [47]. Two types of coatings are being explored: (1) metallic alloys that form durable oxides and (2) ceramics for increased corrosion resistance and as thermal barriers. Alloy films of the general type MCrAlY have been found in gas turbine experience to form high-temperature stable protective oxides that greatly extend the protective life of the alloy coating and protect the substrate turbine blade alloy. CrAlY alloy coatings of nickel, cobalt, and iron have been evaluated [48, 49]. Thermal barrier coatings consist of high-temperature stable ceramics (e.g., ZrO_2) with low thermal conductivity. Such coatings increase the permissible turbine inlet temperature for internally cooled blades by providing a temperature differential of up to 200°C between the outer protective surface and the surface of the structural blade alloy [50]. The degradation of ceramic coatings due to fracture or spalling is an important concern, and the importance of coating microstructure in the control of these failure modes has been reported [51].

2. Encapsulants for Photovoltaics

Encapsulation of photovoltaic cells appears to be essential to inhibit corrosion of the metallic electrode at the semiconductor interface, to prevent the erosive scouring of the semiconductor surface, and to mitigate the effects of dust and hail. Although most encapsulant candidates are thick polymeric or glass films, antireflection coatings on the semiconductor surface, if suitably dense, can also serve a protective function. In particular, gasless ion plating of Al_2O_3 and SiO has shown considerable promise as a technique for providing an impermeable antireflection coating. Essentially, the first several monolayers of dielectric materials are implanted into the semiconductor surface, thus providing an extremely adherent coating.* The electrostatic bonding of thin glass to photovoltaic cells has received considerable attention [52]. In this technique adhesion is achieved at elevated temperatures by applying a dc voltage across the interface to be bonded, thus producing a migration of ions. Integral rf-sputtered borosilicate glass covers (40–60 μm thick) have been evaluated and have performed well except for some moisture-induced delamination [53].

The UV and erosion stability of the solar transmittance of the encapsulation system (outer cover, adhesives, pottant, antireflection coating) over a 20-year cell life is an important concern. Adhesion of the antireflection coating to the front surface of the photovoltaic cell and the ability of the encapsulant film to disperse the impact energy of incident dust and hail

*Work performed under contract to the Jet Propulsion Laboratory by Endurex, A Division of Illinois Tool Works.

particles are also critical, because moisture penetration must be minimized and cracked solar cells may exhibit seriously degraded performance. Reduction of encapsulation cost is also important in producing economical terrestrial devices. The Low Cost Silicon Solar Array Project (LSA) has concentrated significant efforts in this area, and two thorough reviews of the subject are available [54, 55]. Although numerous module configurations and materials combinations exist, leading material candidates for the outer cover are glass, an acrylic (KORAD 201-R produced by XCEL), and a halocarbon (Tedlar UT-BG30 produced by Dupont). The addition of a thin, hard, abrasion-resistant coating to the polymer surface is being investigated.

3. Protective Coatings for Reflectors

Reflectors are unavoidably area-intensive components of solar collectors and represent 25%–50% of the capital cost. It is imperative that corrosive impurities such as chloride and sulfide ions in the presence of moisture not jeopardize adequate reflector lifetime.

The conventional commercial silvered reflector has a reflective bilayer comprised of 1000 Å of silver and 250 Å of copper covered by a thick, protective paint film [21]. This mirror, bonded to a support structure, is the basic consideration for glass heliostats (individual reflecting elements of a Fresnel central receiver solar thermal system). Preliminary evidence suggests that commercial reflectors may not meet the extremely rigorous requirements of reflector performance in current designs for high-concentration solar applications because localized corrosion of the silver layer is observed after short-term field exposure. The source of the impurities responsible for the corrosion has not been identified and may be the protective paint, the adhesives, nearby structural elements, or impurities in the environment. The presence of moisture is an important factor in the degradation, and moisture from frost, dew, and cleaning will be unavoidable even in a desert environment. The solution to this potentially serious problem may require improved quality control of impurities in mirror components or development of an alternate protective scheme, such as improved paints or intermediate impermeable layers.

Innovative second-generation mirror structures may be required to solve simultaneously performance, cost, and durability problems, and a large solar concentrator market could justify the investment in capital-intensive equipment with high throughput. Vapor-phase techniques such as ion plating will require evaluation, and a laminated mirror structure using thin glass ($\sim$ 1 mm) may provide environmental stability at an acceptable cost.

Advanced aluminum and substrate silver mirrors are also receiving attention with the potential for providing lower cost. The natural oxide on aluminum at an equilibrium thickness of less than 50 Å is not sufficient protection for aluminum mirrors, and the oxide is typically augmented by anodization to a thickness of ~ 500 Å [56]. Silazane-based varnishes have been explored for the protection of substrate silver mirrors [57], and silicone resins are presently being evaluated for this application.*

4. Working-Fluid Protective Coatings

Surfaces exposed to working fluids in solar heat engines and receivers must be protected to ensure adequate lifetimes. Titanium and aluminum heat exchangers are being proposed to work between ammonia and seawater in closed Rankine cycle ocean thermal energy conversion systems. Steam at 550°C and 14 MPa (2000 psig), molten salts, and molten aluminum are working fluids in solar thermal systems. In a different realm, low-cost material approaches to water and space heating must evaluate the impact on system lifetime of the corrosion induced by water and antifreeze solutions. In each case it is important that the protective film not impede the flow of heat and thereby reduce system performance.

A molten aluminum heat transfer medium contained within a stainless steel cavity has been proposed for a point-focusing parabolic dish system. Such a system will require a protective layer at the aluminum–stainless steel interface to prevent the diffusion of aluminum into the stainless steel, with eventual failure of the containment. While MoS_2 has been demonstrated to have potential benefit in this application [58], considerable testing will be required to ensure material compatibility in a working environment.

The use of chemical conversion treatments on metal surfaces in contact with fluids and other corrosive environments is a well-established practice. The degree to which such techniques need be modified for solar applications will be determined by the impurities present and the heat flow characteristics desired. As an example, municipal waste pyrolysis creates reactive environments that can lead to very high rates of attack on conventionally protected steel alloys. New protective systems may be needed as the most cost-effective approach in guaranteeing adequate system lifetime (as opposed to prescreening the incoming waste material).

5. Antibiofouling Coatings

Marine surfaces for preventing or at least impeding the buildup of organic deposits on ship hulls, moorings, buoys, and cables have been the

* Work performed under contract to the Jet Propulsion Laboratory and the Solar Energy Research Institute by Dow Corning.

subject of research for over 40 years. Copper–nickel alloys provide good biofouling resistance because of a concentration of copper ions near the surface in the water. Paints containing heavy metals that can be leached at a controlled rate are also effective in this application. However, increased concern for estuaries and fisheries has forced a closer look at the effects of heavy metals as selective poisons in marine applications, and new coatings may be required.

Closed Rankine cycle ocean thermal energy conversion technology uses vast areas ($\sim 5 \times 10^3$ m^2/MW$_e$) of heat exchanger surface between the working fluid (ammonia) and ocean water to convert the 20°C temperature difference between the surface and deep ocean water to electricity. Heat transfer coefficients in excess of 5.7 kW/m^2 °C (1000 Btu/ft^2 °F h) are required for commercial viability. The growth on the heat exchanger walls of even thin organic layers (< 25 μm) from the ocean water begins to reduce the heat transfer coefficient to unacceptable levels. With present shell-and-tube heat exchanger designs the biofouling problem is controlled by brushes or cleaning spheres that scour the surface. Advanced fluted designs may not be so amenable to cleaning, and coatings that protect the surface from rapid buildup are one potential solution to the problem. Since early plants are planned for island and gulf coast sites, the solution will probably have to meet strict environmental regulations.

9.5 SPECIAL SURFACE ENERGIES

The energy associated with a free surface is the sum of two terms: the dispersive (van der Waals) and polar components. Kaelble [59] discusses the physical and chemical origins of the surface energy; however, very little discussion appears in the literature regarding the role of surface energy in solar applications. Such applications include the determination of dust adhesion on reflector and glazing surfaces and the wetting of fluids on heat exchanger walls.

Surface energy may be a very important parameter determining the adhesion of dust. Dust is a significant concern for mirror and glazing performance [60, 61], with up to a 0.50 decrease in solar reflectance or transmittance observed in 1 month. Detailed studies of the adhesion of microscopic dust particles (0.5–5 μm) have demonstrated that a primary adhesion mechanism is the leaching of inorganic salts from the dust particle by locally condensed moisture. When the water evaporates, an insoluble "glue" that may extend for several dust particle diameters remains. This effect is important for solar mirrors even in dry environments such as desert sites, which are particularly prone to dew and frost. It has been shown that a surface with energy of very low polar component, such as sputtered Teflon and other fluorinated compounds, dramatically reduce the effects of the glue [62]. The advantage of reduced dust particle adhesion, however, is partly negated by the tendency of films of such low surface energy to be poorly adherent, easily scratched, and thus difficult to clean. In addition, cleaning solutions do not "wet" the surface well, and thus removal of even poorly adherent dust particles may require specialized treatment. Further basic understanding of the dust–mirror or glazing interface will be required to evaluate the performance of improved coatings or cleaning solutions in removing adhered dust.

Although perhaps more amenable to an engineering solution, turbulent flow at the boundary between a heat exchanger wall and a working fluid may allow dewetting of the surface for low surface energies and a decrease in the heat transfer coefficient. A surface with a very high-polar-energy component may remain covered over a wider flow regime because of the reduced free energy in this condition. Research is warranted into the stability of the surface energy and the magnitude of the thermal effects to be achieved.

9.6 MEMBRANES

Selective membranes are thin films which achieve selective mass transport by a specially engineered microstructure. In a sense membranes are akin to protective films where the constraint of complete impermeability is relaxed to exclude certain molecules or ions. Selectively *permeable* membranes contain holes sufficiently small to block large molecules while readily transmitting smaller molecules such as water. The pores may average less than 10 nm in diameter. *Ion-selective* membranes transport specific ions (or a set of very similar ions) such as H^+, Cl^-, or Na^+, by ion-exchange transport through the film structure.

Selective membranes have been proposed for use in a number of solar technologies. The most important are the redox battery for electric energy storage in photovoltaic systems, the sodium heat engine and sodium–sulfur (Na–S) battery for high-temperature electrical conversion and storage, the reverse electrodialysis (RED) cell for direct electrochemical production of electricity using salt brines, and pressure-retarded osmosis (PRO) systems for electric power production from osmotic head hydroelectric generators which also operate at a salt brine–fresh water interface.

The redox battery [63] and the RED cell [64] use cation-selective and anion-selective membranes. In the redox battery electric energy is stored as an electrochemical potential between chloride solutions containing Cr^{2+} – Cr^{3+} and Fe^{3+} –Fe^{2+} couples, as illustrated in Fig. 9.6a. The geometry of

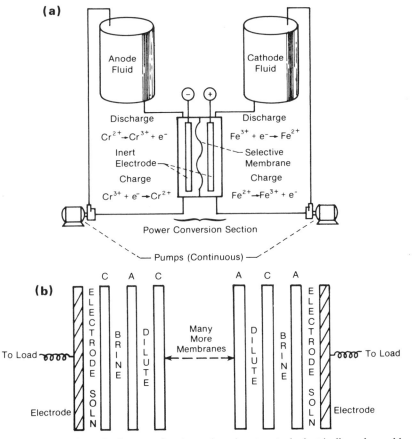

Fig. 9.6 (a) Schematic diagram of an iron–chromium two-tank electrically rechargeable redox flow cell. (From Thaller [69].) (b) Schematic diagram of a RED stack cell. (From Lacey [63].) C—cation-exchange membrane; A—anion-exchange membrane.

a RED electrochemical cell is illustrated in Fig. 9.6b. The RED cell and redox battery systems use ion-selective membranes designed for electrodialysis; these membranes are not well suited for potential solar applications. Ion-selective membranes suffer from high cost, high internal (electrical) resistance, and loss of integrity (mostly pinholes). The cost (approximately $10/m^2) and integrity are important because of the very large membrane areas required (typical RED cell power production is projected to be approximately 5 W/m^2). Thinner, stronger membranes at lower cost per unit area are required. Such membranes could be expected to have higher ionic conductance (lower internal resistance) and greater integrity.

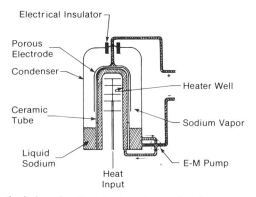

Fig. 9.7 Schematic design of sodium heat engine. A closed container is partially filled with liquid sodium working fluid and is physically divided into high- and low-pressure regions by a pump and a tubular membrane of solid electrolyte which bears a porous metal electrode. The inner section of the device is maintained at a temperature T_2 by a heat source, while the outer section is maintained at a temperature T_1 ($T_2 > T_1$) by a heat sink. The temperature differential between the two regions gives rise to a sodium vapor pressure differential across the solid electrolyte membrane.

During operation sodium travels a closed cycle through the device. Starting in the high-temperature (high-pressure) region, the heat source raises the incoming liquid sodium to temperature T_2. Since β''-alumina has high conductivity for sodium ions and negligible electronic conductivity, a mole of electrons must exit the high-temperature zone for each mole of sodium ions entering the β''-alumina. Sodium ions then migrate through the β''-alumina membrane in response to the pressure differential (gradient of Gibbs free energy). After passing through the external load, the electrons are recombined with sodium ions at the porous electrolyte interface. Neutral sodium then evaporates from the porous electrode at pressure P_1 and temperature T_2, passing through the vapor space to the condenser at temperature T_1. The condensed liquid sodium is then returned to the high-temperature zone by an electromagnetic pump, completing the cycle. (From Hunt *et al.* [66].)

Both the Na–S battery [65] and the sodium heat engine [66] exploit the high conductivity for sodium ions and low electronic conductivity of β''-alumina. Battery and heat engine processes are carried out at temperatures of 300° and 1000°C, respectively. Figure 9.7 illustrates the heat engine concept. β''-alumina membranes have been produced at tube wall thicknesses as low as 100 μm [68]. The long-term resistance of the ceramic to cracking is an unresolved important issue.

Pressure-retarded osmosis systems use cast polymer membranes (Loeb–Sourirajan membranes) or composite membranes designed for use in reverse osmosis desalinization [69]. Both Loeb–Sourirajan and composite membranes are physical membranes that have an active layer approximately 100 nm thick containing pores about 20 nm in diameter. This thin and fragile layer is supported by a much thicker ($\sim$1000 nm) porous layer. Ion concentration in the porous support layer polarizes the PRO membrane and reduces the flux dramatically. The very high pressures (> 10 MPa) involved in PRO cause a gradual compression of present polymer membranes and thus a decrease in their permeability [approximately 30% decrease in flux at 10 MPa (1500 psig) over a period of 36 days]. A stronger semipermeable membrane with a much thinner support layer could improve the performance of PRO membranes by a factor of at least 3. Attempts to use glass have not been successful, because of the rapid dissolution of the glass and the consequent widening of the pores in a brine environment.

9.7 CONCLUSION

This chapter illustrates the breadth of thin-film applications across the spectrum of solar technologies. Photovoltaics and selective solar absorbers have attracted and continue to attract considerable attention, but it is important that other critical thin-film problems be addressed in a timely manner.

REFERENCES*

1. M. P. Thekaekara, *Solar Energy* **14**, 109 (1973).
2. Y. S. Touloukian, "Thermal Radiative Properties," Volume 9, Coatings. IFT/Plenum Press, New York, 1972.
3. K. Masterson (ed.), *Status Report of Solar Optical Materials Planning Committee.* SERI Rep., SERI/TR-31-137 (1979).
4. M. Abrams, The Effectiveness of Spectrally Selective Surfaces for Exposed, High-Temperature Solar Absorbers. Sandia Laboratories Rep., SAND-77-8300 (1978).
5. D. Trotter and A. J. Sievers, *Appl. Phys. Lett.* (to be published).
6. P. J. Call, National Plan for Absorber Surface R&D. SERI Rep., SERI/TR-31-103 (1978).
7. R. E. Hahn and B. O. Seraphin, *in* "Physics of Thin Films," Vol. 10. Academic Press, New York, 1978.
8. B. O. Seraphin and A. B. Meinel, *in* "Optical Properties of Solids, New Developments," Chapter 17. North Holland Publ., Amsterdam, 1978.
9. L. Melamed and G. M. Kaplan, *J. Energy* **1**, 100 (1977).
10. D. M. Mattox, G. J. Kominiak, R. R. Sowell, and R. B. Pettit, *Selective Solar Photothermal Absorbers.* Sandia Laboratories Rep., SAND-75-0361 (1975).
11. D. Grimmer, Solar Selective Absorber Coatings on Glass Substrates. Los Alamos Scientific Laboratories Rep., LA-UR-77-2700 (1977).

*Because of the early development state of many of the solar technologies, much of the pertinent information is contained in government reports rather than the open literature. Government reports are available from NTIS, U.S. Dept. of Commerce, 5285 Port Royal Road, Springfield, Virginia 22161.

12. P. J. Call (ed.), *Proc. DOE/DST Thermal Power Syst. Workshop Selective Absorber Coatings* SERI Rep., SERI/TP-31-061 (1978).
13. Fourth Annual Rep. University of Sydney, Energy Resource Centre, Sydney, Australia (1977).
14. M. van der Leij, "Spectral-Selective Surfaces for the Thermal Conversion of Solar Energy." Delft Univ. Press, 1979.
15. L. R. Gilbert, R. Messier, and R. Roy, *Thin Solid Films* **54**, 149 (1978).
16. J. C. C. Fan, *Thin Solid Films* **54**, 139 (1978).
17. R. R. Sowell and R. B. Pettit, *Plating and Surface Finishing* **65** (No. 10), 42 (1978).
18. A. P. Bradford and G. Haas, *Solar Energy* **9**, 32 (1964).
19. D. M. Trotter, in Summary Report of the Solar Reflective Materials Technology Workshop (M. A. Lind and L. E. Ault, eds.). Battelle Pacific Northwest Laboratory Rep. PNL-2763, 75 (1978).
20. M. A. Lind and L. E. Ault (eds.), Summary Report of the Solar Reflective Materials Technology Workshop, Battelle Pacific Northwest Laboratory Rep., PNL-2763 (1978).
21. B. Schweig, "Mirrors: A Guide to the Manufacture of Mirrors and Reflecting Surfaces." Pellham Books, London, 1976.
22. H. L. Hampton and M. A. Lind, Weathering Characteristics of Potential Solar Reflector Materials: A Survey of the Literature. Battelle Pacific Northwest Laboratories PNL-2824/UC 62 (1978).
23. A. C. Ratzel and R. B. Bannerot, *in Proc. 1977 Flat-Plate Solar Collector Conf.* CONF-77-0253, 387 (1978).
24. F. L. Bouquet, Aging Characteristics of Mirrors for Solar Energy Applications. Jet Propulsion Laboratory Rep., JPL5102-116 (1979).
25. "Modern Plastics Encyclopedia," Vol. 54, No. 104. McGraw Hill, New York, 1977.
26. D. W. Van Krevelen, "Properties of Polymers," p. 524. American Elsevier, New York, 1976.
27. B. O. Seraphin, *Thin Solid Films* **39**, 87 (1976).
28. R. B. Stephens and G. D. Cody, *Thin Solid Films* **45**, 19 (1977).
29. M. J. Minot, *J. Opt. Soc. Am.* **67**, 1046 (1977).
30. O. K. Hussman and K. Kerner, *J. Vac. Sci. Tech.* **14**, 200 (1977).
31. T. H. DiStefano, G. D. Pettit, J. M. Woodall, and J. J. Cuomo, *Appl. Phys. Lett.* **32**, 676 (1978).
32. T. B. Reed, Solid State Research Rep., MIT Lincoln Laboratory, p. 21 (1969).
33. J. C. C. Fan, T. B. Reed, and J. B. Goodenough, *Proc. IECEC, 9th* p. 341 (1974).
34. J. C. C. Fan and F. J. Bachner, *Appl. Opt.* **15**, 1012 (1976).
35. M. N. Koltun and Sh. A. Faiziev, *Geliotekhnika* **13**, 28 (1977).
36. G. Haacke, *Appl. Phys. Lett.* **30**, 380 (1977).
37. C. B. Greenberg, *J. Electrochem. Soc.* **126**, 337 (1976).
38. N. B. Rekant and I. I. Borisova, *Geliotekhnika* **3**, 42 (1966).
39. J. A. Apfel, *J. Vac. Sci. Tech.* **12**, 1016 (1975).
40. J. L. Vossen, Transparent conducting electrodes, *in* "Physics of Thin Films," Chapter 1, Vol. 9. Academic Press, New York, 1977.
41. J. C. C. Fan, F. J. Bachner, and G. H. Foley, *Appl. Phys. Lett.* **31**, 773 (1977).
42. P. A. Iles and S. I. Soclof, *Proc. IEEE Photovoltaics Specialists Conf., 12th* p. 978 (1976).
43. C. R. Wronski, D. E. Carlson, and R. E. Daniel, *Appl. Phys. Lett.* **29**, 602 (1976).
44. W. Anderson, *Proc. ERDA Semiann. Photovoltaics Adv. Mater. Program Rev. Meeting* CONF-77-0318, p. 194 (1977).
45. H. Birnbreier, German Patent 2,454,206 (1976).
46. R. M. Winegarner, *Proc. Ann. Meeting AS/ISES* **1**, Section 26 (1977).
47. G. W. Goward, *Thin Solid Films* **53**, 223 (1978).

48. S. G. Young and G. R. Zellars, *Thin Solid Films* **53**, 241 (1978).
49. J. R. Rairden, *Thin Solid Films* **53**, 251 (1978).
50. C. H. Liebert, *Thin Solid Films* **53**, 235 (1978).
51. P. F. Becher, R. W. Rice, C. Ch. Wu, and R. L. Jones, *Thin Solid Films* **53**, 225 (1978).
52. P. R. Younger, W. S. Kreisman, and A. R. Kirkpatrick, Integral Glass Encapsulation for Solar Arrays. A report prepared for the Jet Propulsion Laboratory by Spire Corp., JPL/954521 (1977).
53. R. L. Crabb and J. C. Larue, *Proc. IEEE Photovoltaics Specialists Conf., 12th* p. 577 (1976).
54. E. F. Cuddihy, Encapsulation Materials Trends Relative to the 1986 Cost Goals. Jet Propulsion Laboratory Rep., JPL-LSA-5101-61 (1978).
55. E. F. Cuddihy, B. Baum, and P. Willis, Low-Cost Encapsulation Materials for Terrestrial Solar Cell Modules. Jet Propulsion Laboratory Rep., JPL-LSA-5101-78 (1978).
56. G. Haas, *J. Opt. Sci. Am.* **39**, 532 (1949).
57. R. A. Zakhidov, A. Ismanzhanov, I. I. Gribelyuk, L. A. Dubrovskii, A. G. Sheinina, and N. N. Baibakova, *Geliotekhnika* **13**, 65 (1977).
58. R. Avazian, Heat Sink Heater. NASA Brief No. MFS-19334 (1977).
59. D. H. Kaelble, "Physical Chemistry Adhesion," p. 149. Wiley (Interscience), New York, 1970.
60. H. P. Garg, *Solar Energy* **15**, 299 (1974).
61. R. S. Berg, Heliostat Dust Buildup and Cleaning Studies. Sandia Laboratories Rep., SAND-78-0510 (1978).
62. R. Williams, RCA Corp., David Sarnoff Research Center, private communication.
63. M. Warshay and L. O. Wright, *J. Electrochem. Soc.* **124** (1977).
64. R. E. Lacey, Energy by Reverse Electrodialysis. A report prepared for the Department of Energy by the Southern Research Institute (1978).
65. Battery Technology—An Assessment of the State of the Art. A report compiled by TRW, Inc. (1978).
66. T. K. Hunt, N. Weber, and T. Cole, *Proc. IECEC, 13th, San Diego, California* (1978).
67. G. J. Tennenhouse and R. A. Pett, Fabrication of Thin Layer Beta Alumina. A Contract Report, NASA-CR-135-308 (1977).
68. R. S. Norman, *Science* **186**, 350 (1974).
69. L. H. Thaller, Electrically Rechargeable Redox Flow Cells. NASA Technical Memorandum, NASA TM X-71540 (1974).

Bibliography
of Thin Film
Books

1 PHYSICS, CHEMISTRY, AND PROCESSES
IN THIN FILMS

1. L. Holland, "Vacuum Deposition of Thin Films." Wiley, New York, 1956.
2. H. Mayer, "Structure and Properties of Thin Films." Wiley, New York, 1959.
3. A. Vasicek, "Optics of Thin Films." North Holland Publ., Amsterdam, 1960.
4. C. A. Neugebauer, J. B. Newkirk, and D. A. Vermilyea (eds.), "Structure and Properties of Thin Films." Wiley, New York, 1962.
5. H. G. F. Wilsdorf (ed.), "Thin Films." American Society for Metals, Metals Park, Ohio, 1963.
6. M. Francombe and H. Sado (eds.), "Single Crystal Films." Pergamon, Oxford, 1964.
7. E. Passaglia, R. R. Stromberg, and J. Kruger (eds.), "Ellipsometry in the Measurement of Surfaces and Thin Films," Misc. Publ. 256. National Bureau of Standards, Washington, D. C., 1964.
8. M. Prutton, "Thin Ferromagnetic Films." Butterworth, London, 1964.
9. O. S. Heavens, "Optical Properties of Thin Solid Films." Dover, New York, 1965.
10. R. F. Soohod, "Magnetic Thin Films." Harper, New York, 1965.
11. R. Niedermayer and H. Mayer (eds.), "Basic Problems in Thin Film Physics." Vandenhoeck-Ruprecht, Gottingen, 1966.
12. D. R. Lamb, "Electrical Conduction in Thin Insulating Films." Methuen, London, 1967.
13. B. Schwartz and N. Schwartz (eds.), "Measurement Techniques for Thin Films." Electrochemical Society, New York, 1967.
14. N. N. Axelrod (ed.), "Optical Properties of Dielectric Films." Electrochemical Society, New York, 1968.
15. K. L. Chopra, "Thin Film Phenomena." McGraw-Hill, New York, 1969. Reprint, 1979.
16. R. W. Berry, D. M. Hale, and M. T. Harris, "Thin Film Technology." Van Nostrand-Reinhold, Princeton, New Jersey, 1970.
17. O. S. Heavens, "Thin Film Physics." Methuen, London, 1970.

18. L. I. Maissel and R. Glang (eds.), "Handbook of Thin Film Technology." McGraw-Hill, New York, 1970.
19. J. C. Anderson (ed.), "Chemisorption and Reactions on Metallic Films," Vols. 1 and 2. Academic Press, New York, 1971.
20. K. D. Leaver and B. N. Chapman, "Thin Films." Crane Russak Co., New York, 1971.
21. J. G. Simmons, "DC Conduction in Thin Films." M & B Monograph, New York, 1971.
22. B. N. Chapman and J. C. Anderson (eds.), "Science and Technology of Surface Coatings." Academic Press, New York, 1973.
23. G. Hass (ed.), "Physics of Thin Films: Advances in Research and Development," Vols. 1–10. Academic Press, New York, 1973–1978.
24. L. I. Maissel and M. H. Francombe, "An Introduction to Thin Films." Gordon and Breach, New York, 1973.
25. T. I. Coutts, "Electrical Conduction in Thin Metal Films." Elsevier, Amsterdam, 1974.
26. J. G. Dash, "Films on Solid Surfaces: The Physics and Chemistry of Physical Absorption." Academic Press, New York, 1975.
27. C. H. S. Dupuy and A. Cachard (eds.), "Physics of Nonmetallic Thin Films." Plenum Press, New York, 1976.
28. Z. Knittl, "Optics of Thin Films: An Optional Multilayer Theory." Wiley (Interscience), New York, 1976.
29. H. Kressel (ed.), "Characterization of Epitaxial Semiconductor Films." Elsevier, Amsterdam, 1976.
30. L. Eckertova, "Physics of Thin Films." Plenum Press, New York, 1977.
31. J. E. E. Baglin and J. M. Poate (eds.), "Thin Film Phenomena—Interfaces and Interactions." Electrochemical Society, New Jersey, 1978.
32. J. M. Poate, K. N. Tu, and J. W. Mayer (eds.), "Thin Films: Interdiffusion and Reactions." Wiley (Interscience), New York, 1978.
33. J. Yarwood (ed.), "Vacuum and Thin Film Technology." Pergamon, Oxford, 1978.
34. J. L. Vossen and W. Kern (eds.), "Thin Film Processes." Academic Press, New York, 1979.

2 APPLICATIONS OF THIN FILMS

1. L. Holland (ed.), "Thin Film Microelectronics." Chapman and Hall, London, 1965.
2. J. C. Anderson (ed.), "The Use of Thin Films in Physical Investigations." Academic Press, New York, 1965.
3. R. A. Coombe (ed.), "The Electrical Properties and Applications of Thin Films." Pitman Publ., New York, 1967.
4. H. A. Macleod, "Thin Film Optical Filters." Elsevier, Amsterdam, 1969.
5. A. C. Tickel, "Thin Film Transistors." Wiley (Interscience), New York, 1969.
6. Z. H. Meiksin, "Thin and Thick Films for Hybrid Microelectronics." Lexington Books, Lexington, Massachusetts, 1976.
7. T. J. Coutts (ed.), "Active and Passive Thin Film Devices." Academic Press, New York, 1978.
8. C. E. Jennett, "The Engineering of Microelectronic Thin and Thick Films." Macmillan, New York, 1978.

3 SPECIFIC THIN-FILM MATERIALS

1. L. Young, "Anodic Oxide Films." Academic Press, New York, 1961.
2. J. W. Diggle, "Oxides and Oxide Films." Dekker, New York, 1972.
3. J. G. Daunt and E. Lerner (eds.), "Monolayer and Submonolayer Helium Films." Plenum Press, New York, 1973.
4. W. D. Westwood, N. Waterhouse, and P. Wilcox, "Tantalum Thin Films." Academic Press, New York, 1975.
5. A. G. Stanley, "Cadmium Sulfide." Publ. 77-78, Vols. 1 and 2. Jet Propulsion Lab, Pasadena, California, 1978.

4 THIN-FILM BIBLIOGRAPHY

1. H. Mayer, "Physics of Thin Films: Complete Bibliography," Vol. 1 and 2. International Publ., New York, 1972.

Index